THE MOON
IN THE
POST-APOLLO ERA

by

ZDENĚK KOPAL

University of Manchester, England

SPRINGER-SCIENCE+BUSINESS MEDIA, B.V.

Library of Congress Cataloging in Publication Data

Kopal, Zdeněk, 1914–
    The Moon in the post-Apollo era.

    (Geophysics and astrophysics monographs ; v. 7)
    Includes bibliographies and index.
    1. Moon. I. Title. II. Series.

QB581.K595   559.9'1   74–26877
ISBN 978-90-277-0278-4   ISBN 978-94-010-2101-2 (eBook)
DOI 10.1007/978-94-010-2101-2

Published by D. Reidel Publishing Company,
P.O. Box 17, Dordrecht, Holland

Sold and distributed in the U.S.A., Canada, and Mexico
by D. Reidel Publishing Company, Inc.
306 Dartmouth Street, Boston,
Mass. 02116, U.S.A.

THE MOON IN THE POST-APOLLO ERA

GEOPHYSICS AND ASTROPHYSICS MONOGRAPHS

AN INTERNATIONAL SERIES OF FUNDAMENTAL TEXTBOOKS

VOLUME 7

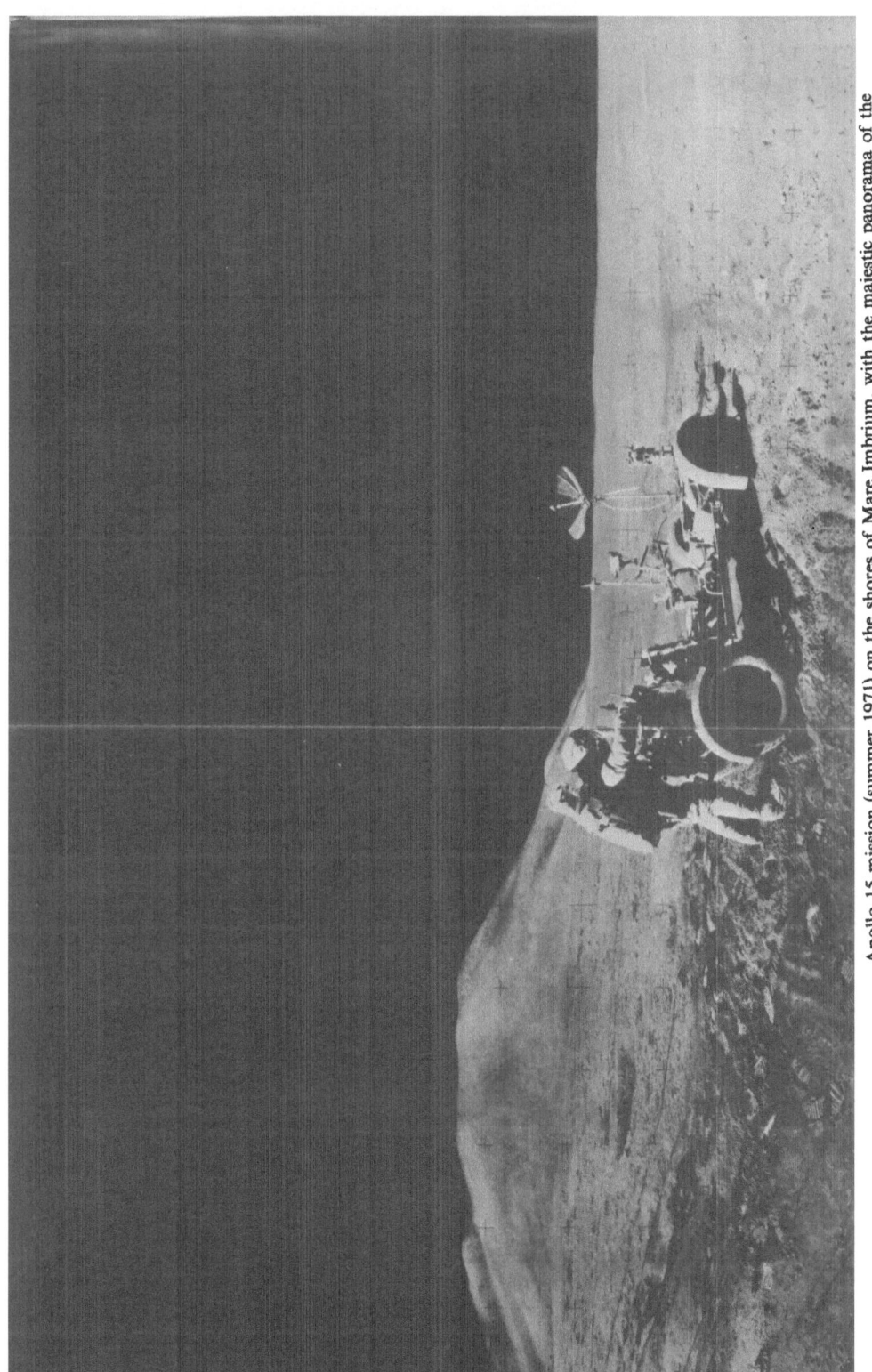

Apollo 15 mission (summer 1971) on the shores of Mare Imbrium, with the majestic panorama of the lunar Apennines in the background. (By courtesy of NASA and ACIC)

TABLE OF CONTENTS

PREFACE

The aim of the present book will be to summarize the results of the space exploration of the Moon in the past fifteen years – culminating in the manned Apollo missions of 1969–1972 – on the background of our previous acquaintance with our satellite made in the past by astronomical observations at a distance.

Astronomy is one of the oldest branches of science conceived by the inquisitive human mind; though until quite recently it had been debarred from the status of a genuine experimental science by the remoteness of the objects of its study. With the sole exception of meteoritic matter which occasionally finds its way into our laboratories, all celestial bodies could be investigated only at a distance: namely, from the effects of attraction exerted by their mass, or from the ciphered messages of their light carried by nimble-footed photons across the intervening gaps of space.

A dramatic emergence of long-range spacecraft – capable of carrying men with their instruments not only outside the confines of our atmosphere, but to the actual surface of our nearest celestial neighbour – has since 1957 thoroughly changed this time-honoured picture. In particular (as we shall detail in Chapter 1 of this book) space astronomy of the Moon is barely 15 years old. But relative infant as it is by age, it has already provided us with such a tremendous amount of new and previously inaccessible scientific data as to virtually revolutionize our subject. The aim of this book will be to provide their digest which could be read with understanding – and, I hope, interest – on the part of the more general reader.

It is not for the first time that the present writer addressed himself to this topic in recent years. In particular, his *Introduction to the Study of the Moon* (D. Reidel Publ. Company, Dordrecht, 1966) went through the press when the first Surveyor landed on the Moon, and is now out of date as well as out of print. Its second edition of 1969 (hereafter referred to as the *Moon II*) is, however, still available, and contains much astronomical background information which need not be repeated in this book. On the other hand, only most cursory results of the first Apollo landing could have been mentioned in its text at page-proof stage; and in this respect its contents are likewise out of date. To make good this shortcoming, in the present volume attention will be focussed primarily on the results of the lunar manned missions completed between 1969–1972. This is the first book in which this will be done; and our aim will be to incorporate these results with all our previous knowledge into a coherent picture of astronomy, physics, chemistry as well as geology or mineralogy of our only natural satellite.

At first sight, the difficulties inherent in such a task appear formidable. Where so many perplexing questions from different scientific disciplines are to be brought into the common focus, no single expert can speak *ex cathedra* to all. Brash young Alexanders of the past are known to have tackled some of our knottiest problems by brutal force. The present writer cannot – alas – compete in such gymnastics, if alone because of advancing years; and cannot, likewise, embark on a synthesis of relevant facts and reasonable theories otherwise than by availing himself – at least partly – of second-hand knowledge.

To amass this knowledge proves by itself a matter of some difficulty. The volume of new data acquired in the field of lunar research since 1969 is simply tremendous – seldom in the annals of science do we encounter a comparable 'explosion of information'. To give some examples, the published Proceedings of the four annual Houston Lunar Science Conferences, alone exceeded 10000 printed pages of original contributions; and the *Moon* journal published since 1969 over 4000 pages of additional literature. To integrate all this information into a single coherent picture constitutes a task which will take many years for its accomplishment. The present little book constitutes the first limited attempt at such an integration – made by an astronomer for whom the Moon is only a part of the solar system. It is, moreover, almost inevitable that not all workers in other fields concerned with lunar studies will agree with the choice of the facts presented in a volume of this size, let alone with their interpretation. But this is only natural; and the whole book was indeed written to this end; for it is only by sustained efforts to gauge the full meaning of all new information which was so suddenly thrust upon us that its implications, not only for the Moon and its past, but also for the origin of the entire solar system can eventually be caught in our grasp.

CHAPTER 1

EXPLORATION OF THE MOON BY SPACECRAFT

If under the term of 'space astronomy' we understand investigations which must be carried out above the main mass of the terrestrial atmosphere, then space astronomy of the Moon is barely 15 years old. Before giving a brief account of its development and accomplishments during this time, however, let us first answer the question: why do we need to investigate the Moon from space?

In lifting instruments for any kind of astronomical observations from the ground to above the main atmospheric air mass (in practice, to 200 km or higher), we aim at freeing ourselves from two limitations imposed by it: namely, its absorption (which prohibits transmission to ground of a very large part of the entire spectrum of extra-terrestrial light sources); and its turbulence (which causes the index of refraction to fluctuate along the line of sight, and thus 'spoils our seeing'). For many types of astronomical work both these obstacles are of comparable importance; but in lunar studies the second one is of paramount significance.

The atmospheric absorption itself restricts lunar ground-based work but little; for most part of the sunlight scattered by the Moon passes through the 'optical window' of transmission of our atmosphere between 3000–10000 Å without much change. The thermal radiation of the Moon itself occurs mainly in the near infrared. However, its maximum emission at daylight temperatures happens to fall within the 8–12 μ infrared window of fair atmospheric transparency, so that its direct measurements are again possible from the ground; and only at lunar nighttime does lunar thermal emission migrate towards longer wavelengths which no longer penetrate the atmosphere. In the transition region of near infrared between wavelengths of 1–5 μ, the spectrum of the Moon has recently been traced with the aid of balloons (cf. Wattson and Danielson, 1965). On the other hand, in the ultraviolet (below the limit of atmospheric ozone absorption) the lunar reflectivity becomes again so low (cf. Lebedinsky *et al.*, 1967a, b) that not much can be learned from observations in this domain even at altitudes from which such observations can be made.

While, therefore, terrestrial atmospheric absorption does not restrict ground-based lunar work in any essential respect, the atmospheric turbulence causing 'bad seeing' is a very different matter. In actual fact, the optical effects of ever-present turbulence seldom permit ground-based telescopes with apertures in excess of 20 inches to attain the resolution permissibly by geometrical optics; and the smallest details on the lunar surface actually discernible by our terrestrial telescopes from the distance of the Earth are seldom less than half a kilometre in size.

For the first 200 years of lunar astronomy, atmospheric degradation of images was of but secondary concern to the observers wrestling with the optical and mechanical imperfections of their instruments. Technology marched on, however, and the stage at which the 'atmospheric seeing' became the limiting factor was reached some time around the turn of this century. It set off an exodus of the old astronomical observatories away from large cities – an exodus which by now has almost been completed – but by itself this move proved to be a mere temporary expedient. By now – in the

Fig. 1.1.   A schematic view of the Russian space station (Luna 3) which photographed for the first time the far side of the Moon on 7 October 1959 (U.S.S.R. Acad. Sci. photograph).

last third of the 20th century – the obvious way of escaping all limitations imposed on ground-based astronomy by an atmosphere overhead is upwards – by launching our instruments into circumterrestrial or heliocentric orbits.

As far as the lunar research is concerned, two courses remain open: either to conduct our observations with the aid of large telescopes in circumterrestrial orbits – of sufficient aperture to attain the requisite ground resolution on the lunar surface – or to send small telescopes or cameras on space missions which would take them much closer to their targets; so that proximity to the object of their study would make up for limited apertures of the optics which they can employ. No large telescope of sufficient resolving power has, however, been injected into circumterrestrial orbit up to this time (1974); so that we had to rely on the second approach so far. The main aim of this chapter will be to describe briefly the type and equipment of the spacecraft deployed for lunar studies in the past 15 years.

Efforts to launch spacecraft to the Moon go back to 1958 – the year following the launch of the first Sputnik – when the American rocket Pioneer 1 (launched on 11 October of that year) reached a distance of 120000 km from the Earth en route to the Moon; and Pioneer 3 (launched on 6 December 1958) attained only a slightly smaller range of 102300 km. It was, however, the Russian Luna 1 (launched on 2 January 1959) which first 'made it' all the way and bypassed the Moon at the smallest distance of only 5965 km (i.e., 3.44 times the Moon's radius of 1738 km) to become the first artificial planet of the Sun launched by the human hand. Moreover, the Russian Luna 2 (launched on 11 September 1959) made history by scoring the first hard impact on the Moon on 13 September – a memorable date in the history of mankind – after a flight $63\frac{1}{2}$ hours. Less than a month later, the Soviet Luna 3 (cf. Figure 1.1), launched on 4 October into a highly eccentric Earth orbit, actually circumnavigated the Moon and unveiled for us the first glimpse of the principal topographic features of its far side (see Figure 1.2).

The following decade between 1960–1970 proved to be very full of accomplishment, and provided the foundations on which the lunar science rests at the present time. More complete data on these missions are listed in the accompanying Table 1-1, in which the entire material concerning 'expendable' missions has been separated into four groups: namely, the lunar (1) fly by's, (2) hard-landers, (3) soft-landers, and (4) orbiters. In what follows, we wish to describe briefly the instrumental equipment and principal accomplishments of these spacecraft in turn.

1. Fly-By Spacecraft

The best-known member of this group – Luna 3 of 1959 (see Figure 1.1) – was dynamically an artificial satellite of the Earth, revolving around the latter in a highly eccentric orbit of period initially close to 16.2 days, and reaching an apogee distance of 469000 km. It ceased to exist on 20 April 1960 (when it was burnt up by air resistance in the terrestrial atmosphere); but all other members of this group have gone into heliocentric orbits of virtually unlimited lifetime. This includes the Russian

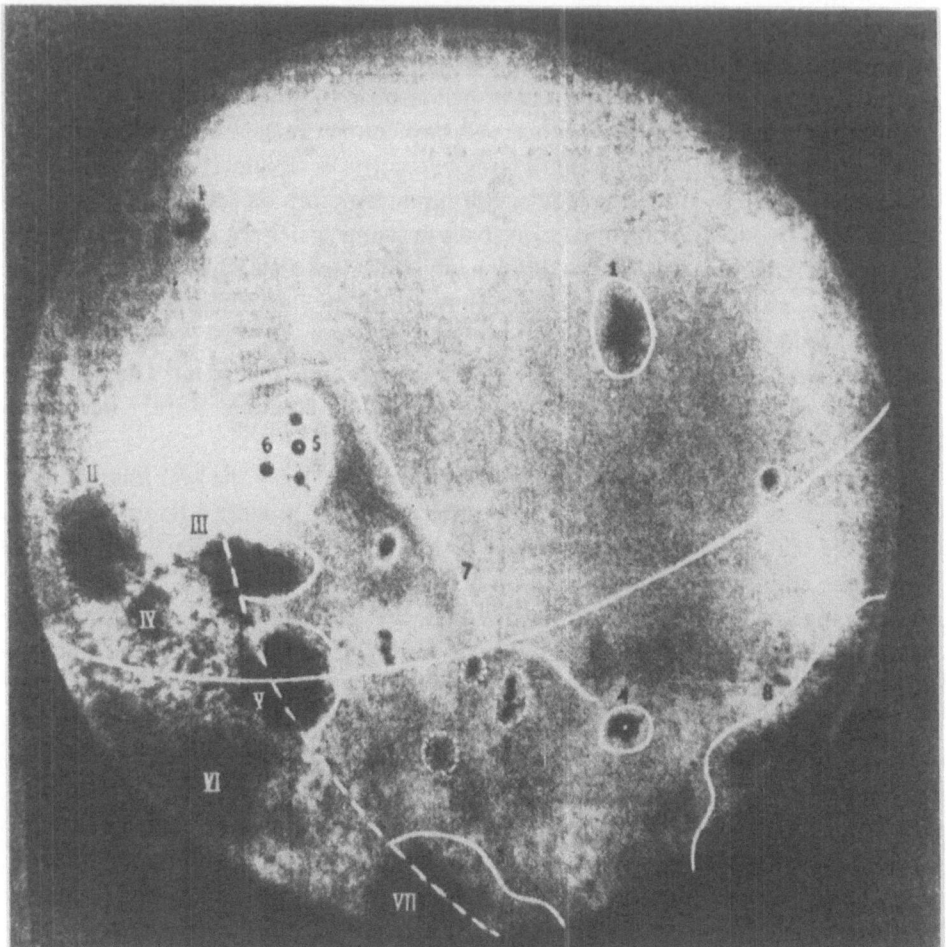

Fig. 1.2. A photograph of the far side of the Moon, secured by Luna 3 on 7 October 1959 at a distance of 66000 km from its target, and telemetered subsequently to the Earth. The solid arc marks the position of the lunar equator; the dotted arc, the limits of the hemisphere visible from the Earth. The Roman numerals indicate the position of the individual maria (I: Mare Humboldtianum; II: Mare Crisium; III: Mare Marginis; IV: Mare Undarum; V: Mare Smythii; VI: Mare Foecunditatis; VII: Mare Australe); while the Arabic numerals indicate the positions of the craters and of certain other formations (1: Mare Moscoviense; 4: the crater Tsiolkovski; 5: Lomonosov; 6: Joliot-Curie; 8: Mare Ingenii). Reproduced from *The First Photographs of the Far Side of the Moon*, U.S.S.R. Acad. Sci., Moscow 1960.

Lunas 4 and 6 (probable attempts at soft landings), or Zond 3 of 1965 which was a practice shot for Mars.

The principal contributions to lunar studies of fly-by spacecraft have been to unveil for us – in 1959 and again in 1965 – the first features of the far side of the Moon, inaccessible to direct observation from the Earth ever since the Moon's axial rotation has become synchronised with its revolution as a result of tidal friction continuously operative in this system. It is true that the 'optical librations' (cf. Chapter 3 of *Moon II*;

TABLE 1-1
List of lunar spacecraft (1959–1972)

Name of spacecraft	Origin	Date of launching	Weight (in kg)*
		(a) *Fly-by*	
Luna 1	U.S.S.R.	1959 January 2	362
Luna 3	U.S.S.R.	1959 October 4	435
Ranger 3	U.S.A.	1962 January 26	330
Ranger 5	U.S.A.	1962 October 18	342
Luna 4	U.S.S.R.	1963 April 2	1422
Luna 6	U.S.S.R.	1965 June 8	1442
Zond 3	U.S.S.R.	1965 July 18	960
		(b) *Hard-landing*	
Luna 2	U.S.S.R.	1959 September 11	390
Ranger 4	U.S.A.	1962 April 23	331
Ranger 6	U.S.A.	1964 January 30	365
Ranger 7	U.S.A.	1964 July 28	366
Ranger 8	U.S.A.	1965 February 17	367
Ranger 9	U.S.A.	1965 March 21	367
Luna 5	U.S.S.R.	1965 May 9	1476
Luna 7	U.S.S.R.	1965 October 4	1506
Luna 8	U.S.S.R.	1965 December 3	1552
Surveyor 2	U.S.A.	1966 September 20	990
Surveyor 4	U.S.A.	1967 July 14	1038
		(c) *Soft-landing*	
Luna 9	U.S.S.R.	1966 January 31	1583 (100)
Surveyor 1	U.S.A.	1966 May 30	990 (292)
Luna 13	U.S.S.R.	1966 December 21	1580 (100?)
Surveyor 3	U.S.A.	1967 April 17	1040 (302)
Surveyor 5	U.S.A.	1967 September 8	1006 (303)
Surveyor 6	U.S.A.	1967 November 7	1008 (303)
Surveyor 7	U.S.A.	1968 January 7	1010 (305)
Luna 17	U.S.S.R.	1970 November 10	(755)
Luna 21	U.S.S.R.	1973 January 8	(840)
		(d) *Orbiting*	
Luna 10	U.S.S.R.	1966 March 31	245
Orbiter 1	U.S.A.	1966 August 10	387
Luna 11	U.S.S.R.	1966 August 24	1640
Luna 12	U.S.S.R.	1966 October 22	?
Orbiter 2	U.S.A.	1966 November 6	390
Orbiter 3	U.S.A.	1967 February 5	385
Orbiter 4	U.S.A.	1967 May 4	390
Explorer 35	U.S.A.	1967 July 19	104
Orbiter 5	U.S.A.	1967 August 1	390
Luna 14	U.S.S.R.	1968 April 7	?
Luna 15	U.S.S.R.	1969 July 13	?
Luna 18	U.S.S.R.	1971 September 2	?
Luna 19	U.S.S.R.	1971 September 28	?
		(e) *Re-entrant*	
Zond 5	U.S.S.R.	1968 September 15–21	Unmanned
Zond 6	U.S.S.R.	1968 November 10–17	Unmanned
Apollo 8	U.S.A.	1968 December 21–27	Manned

Table 1-1 (continued)

Name of spacecraft	Origin	Date of launching	Weight (in kg)*
Apollo 10	U.S.A.	1969 May 18–26	Manned
Apollo 11	U.S.A.	1969 August 9–15	Manned**
Zond 7	U.S.S.R.	1969 August 9–15	Unmanned
Apollo 12	U.S.A.	1969 November 14–24	Manned**
Apollo 13	U.S.A.	1970 April 11–17	Manned
Luna 16	U.S.S.R.	1970 September 12–24	Unmanned
Apollo 14	U.S.A.	1971 January 31– February 9	Manned**
Apollo 15	U.S.A.	1971 July 26–August 7	Manned**
Luna 20	U.S.S.R.	1972 February 14–25	Unmanned
Apollo 16	U.S.A.	1972 April 16–27	Manned**
Apollo 17	U.S.A.	1972 December 7–19	Manned**

* The weights in parentheses given for the soft-landers refer to those of the instrumented packages actually deposited on the lunar surface.
** Manned landings on the surface.

or Kopal and Carder, 1974) of our satellite (due to the eccentricity and inclination of its relative orbit) enable us to see appreciably more than one-half (approximately 59%) of the entire lunar surface from the Earth at one time or another; only 41% being permanently invisible. This was so until October 1959, when the cameras aboard the Russian spacecraft Luna 3 unveiled for us the main features of a major part (50%) of the far side of the Moon (cf. Figure 1.2). The relative position of Luna 3 at the time of photography was such that about 13% of the lunar far side was invisible from the spacecraft, and remained uncharted until 20 July 1965, when another Russian space-probe – Zond 3 – succeeded in recording all but a small fraction of the remainder which was not completed filled till in 1967 by the American Orbiters.

Luna 3 was already an elaborate spacecraft, more than 1 m in size and 435 kg in weight; and the photographic experiment performed aboard it on 7 October 1959 proved to be a forerunner of Zond 3 in 1965 and of the American Orbiters in 1966–1967. The optics of Luna 3 consisted of two lenses 36 and 52.5 mm in diameter, and 200 and 500 mm focal length; working at focal ratios $f/5.6$ and $f/9.5$, respectively. At the time of the photography (between 4 h 30 min and 5 h 10 min UT, 7 October 1959) the spacecraft was between 65 and 68 thousand km behind the far side of the Moon, which would have appeared from this vantage point as a disc of angular diameter close to 3°. The diameter of its image in the focal planes of the two lenses were 10 mm and 25 mm, respectively; and were recorded photographically on 35 mm film with different exposures (of the order of 0.01 s). The exposed film was automatically developed and processed aboard Luna 3, and scanned electronically for telemetry to Earth around the time of the spacecraft's next perigee passage several days later. The actual number of frames taken by each lens during the 40 min of photography was not specified by the investigators; but an atlas subsequently published on the basis of this work contains 30 plates.

The resolution of the lunar ground attained on these photographs was between

25–30 km; and although not all details suggested by them have been confirmed by subsequent work, their principal contribution has been the discovery that the Moon's far side consists predominantly of mountainous ground and contains very few maria. Perhaps nothing illustrates the rapidity of current advance of new knowledge in the field of lunar studies more eloquently than the progress in our acquaintance with the topography of its far side. While the Russian photographs of 7 October 1959 possessed a surface resolution close to that of the early telescopes of Galileo Galilei, six years later – in July 1965 – another Russian space probe, Zond 3, increased this resolution by a factor of ten; and photographs such as that reproduced on Figure 1.4 showed surface details less than 3 km in size – a limit which the American Orbiters a year later diminished to formations 100 m in size with their wide-angle cameras, and close to 1 m with high-resolution optics.

The instrumentation and accomplishments of the Orbiters will be reviewed in their turn in Section 4 of this chapter; but this is the place to describe more fully Zond 3 (cf. Figure 1.3). This 960 kg spacecraft was launched on 18 July 1965 along a trajectory which a close approach to the Moon altered into a heliocentric orbit; and its photographic mission was accomplished on 20 July during the lunar fly-by, when the spacecraft approached the lunar surface at a distance between 9200–9900 km. The optics of Zond 3 consisted of a telephoto lens of only 1.3 cm free aperture and 10.6 cm focal distance, operating at a focal ratio $f/8$, with exposures between 3 and 10 ms taken every $2\frac{1}{4}$ min; and a total of 25 frames were obtained in the course of a 68 min camera run. Like with Luna 3, the films were processed aboard the space station, scanned in 1100 lines per frame, and transmitted to the ground on 29 July when Zond 3 was already more than 2 200 000 km from the Earth. An example of the evidence so acquired is shown on the accompanying Figure 1.4.

The principal contribution to lunar topography furnished by Zond 3 was the photographic coverage of the lunar surface in the region between 110° and 150° of lunar latitude, which was not visible from Luna 3 in 1959, and which was not photographed again till August 1967 by the U.S. Orbiter 5. But the photographic cameras with their associated scanning and transmission facilities were not the only scientific equipment aboard Zond 3. A complement of equal importance was an infrared and ultraviolet diffraction-grating spectrometer, designed to obtain information on the spectral characteristics of moonlight in the domain of wavelengths which do not penetrate through the terrestrial atmosphere. The optical design of the infrared spectrometer was the same as previously employed on Mars 1 space probe. It utilised germanium optics, designed for operation in the wavelength domain between 2.4–4.1 μ, with a spectral resolution of 0.07–0.10 μ. The radiation detector was a lead-sulphide cell, cooled by radiation into space. The spectrometer had four sensitivity levels at different parts of the spectrum; and each scan was accomplished in 20 s.

The actual observations aboard Zond 3 were carried out likewise on 20 July 1965, and a total of 37 infrared spectra recorded at each of the four sensitivity levels in about 50 min. During this time, Zond 3 was about 10 000 km away from the lunar

Fig. 1.3.   The U.S.S.R. Zond 3 spacecraft – a practice shot for Mars – which flew by the Moon on 20 July 1965 and secured photographs of its far side such as reproduced on Figure 1.4 (U.S.S.R. Acad. Sci. photograph).

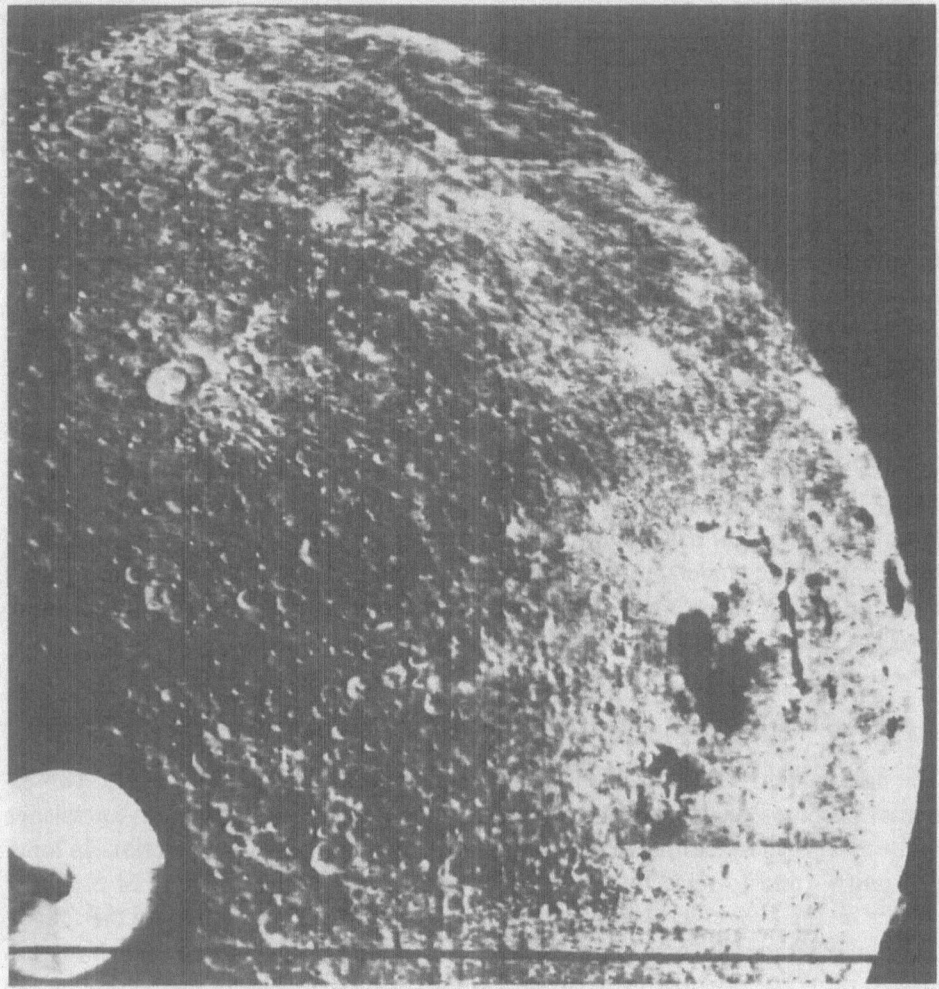

Fig. 1.4.   A photograph of a part of the far side of the Moon obtained by Zond 3 on 20 July 1965. The large dark spot on the right is Mare Orientale (which can be seen in greater detail on Figure 1.29); and the twin crater near the left margin of the field is Vavilov (for greater detail, cf. Figure 2.18). The bright circular disc in the lower left corner represents a photometric scale (U.S.S.R. Acad. Sci. photograph).

surface; therefore, from the space station the Moon appeared as a disc of angular diameter close to 18°; and the spectrometric field of view $7 \times 60'$ corresponded to a surface area of $10 \times 170 \text{ km}^2$.

The results of these measurements were telemetered to the Earth; and their reductions revealed that, in the near infrared (between 3–4 μ), the reflectance of the lunar surface in the continental regions proves to be between five and six times as large as in optical frequencies (0.3–0.6 μ). This is considerably more than Wattson and Danielson inferred the same year from an analysis of Stratoscope II records obtained at an altitude of some 24 km; and the disparity indicates that the atmospheric absorption in the infrared at 24 km may be greater than anticipated.

In the ultraviolet, a different type of photoelectric spectrometer was used to obtain 14 spectrograms of the light scattered from the lunar surface in the domain between 1900–2750 Å (with 14 Å spectral resolution). At the time of the observations, the field of view of the spectrometer was about 3°, subtending on the lunar surface an area close to 500 km across; hence, each scan analysed the light averaged over about one-quarter of the apparent lunar disc (mainly continental areas).

An analysis of these spectra led to a realisation that, in the spectral domain under investigation, the mean albedo of the Moon ranges between 0.010–0.015 – i.e., is about five to six times smaller than in the visible light. In the far ultraviolet the Moon proves, therefore, to be a very poor reflector; and closely approximates a really black body.

The Russian experimenters with Zond 3 discovered, moreover, one additional and very interesting feature in the Moon's UV spectrum: namely, a conspicuous increase in intensity between 2420–2720 Å, present in all spectra, but fluctuating in time between 10–50% of the intensity of the adjacent continuum. Lebedinsky and his collaborators (1967a, b) have tentatively ascribed this phenomenon to luminescence of the lunar surface under fluctuating external influences; and pointed out that, inasmuch as their scans averaged light over large areas of the Moon, the actual luminescence – if localised – may have been very much more intense.

2. Hard-Landing Spacecraft

The hard-landers launched to the Moon in the years of 1959–1965 fulfilled two essential purposes: they enabled us to improve considerably our previous knowledge of the mass of the Moon – essential for all subsequent astronautic operations in lunar proximity – and to relay by television close-up views of the structure of the lunar surface on the 1-m scale, as seen from the spacecraft in the last minutes of their flight. The improvement in our knowledge of the Moon's mass and the Earth: Moon mass-ratio made possible by accurate Doppler tracking of hard-landing (or fly-by) spacecraft in lunar proximity will be detailed later in Table 3-1 of Chapter 3; and for their contributions to our knowledge of the size and shape of the Moon cf. Table 4-1 of Chapter 4.

The principal contributions of the hard-landing U.S. Rangers 7–9 in 1964–1965 were, however, close-up television records of the lunar surface which attained a resolution more than 1000 times as large as that obtainable by the best ground-based telescopes before; and revealed details a metre or less in size. To this end, the last block of this spacecraft carried a complement of six cameras (see Figure 1.5), three of which had 1-in. lenses working at a focal ratio $f/1$; and three others 1.5-in. lenses of 75 mm focal length. The former imaged a field 25° across; the other 8.4°; in addition, others (P_1–P_4 cameras) provided smaller images 2.1°–6.3° in size. The A and B cameras (imaging a field of view of 25° and 8.4°, respectively) produced frames recorded by vidicon TV-tubes and scanned in 1150 lines per picture; their exposures lasted only 2–5 ms (to arrest the motion of the spacecraft); but transmission (at 960

Fig. 1.5. The U.S. Ranger spacecraft (with its solar panels folded in a position for launch) in the shops of the Jet Propulsion Laboratory, California Institute of Technology, where it was designed and built. The cylinder on the top is the omni-directional antenna; and the inset (above, left) shows the location of its optics (photograph reproduced by courtesy of JPL, Calif. Tech.).

Fig. 1.6. A terrestrial photograph of the craters Ptolemaeus, Alphonsus, and Arzachel near the centre of the apparent lunar disc, taken on 22 July 1966 with the 24-in. refractor of the Observatoire du Pic-du-Midi (Manchester Lunar Programme).

Mc s^{-1}) occupied 2.56 s of real time (and reached the Earth with a time-lag of an additional 1.28 s). The smaller images formed by the P-cameras were scanned in 300 lines; and their transmission took just 0.84 s; the energy required for transmission was 60 W. The total weight of the spacecraft was 366 kg, of which optics and telemetry accounted for 172 kg.

The time of flight of the Rangers to the Moon was close to 65 hr; but their actual picture-taking mission occupied only the last quarter of an hour. The transmission commenced at an altitude of 2100–2500 km above the lunar surface, and ended at 600–800 m above it; so that the remaining time of flight was sufficient to transmit only a part of the last P-frames. Altogether, Ranger 7 sent down 4308 individual pictures; Ranger 8, 7137; and Ranger 9, 5814; so that their contributions combined amounted to 17 259 individual frames.

What did this radically new evidence reveal to astronomers at the receiving end of the TV-transmission link? With a ground resolution which exceeded that of the best previous terrestrial work by a factor of 1000, it was possible to extend (at least over limited areas) counts of the lunar craters to formations three orders of magnitude smaller than any we had known thus far, and vastly more numerous. Their statistics revealed for the first time the overwhelming abundance of secondary (and tertiary) craters, formed by impact of, not primary cosmic intruders, but of boulders thrown out from other parts of the Moon by primary cosmic impacts.

In order to place the advances achieved by the Rangers in our acquaintance with the lunar surface in proper perspective, attention is invited to the accompanying Figure 1.6, which shows one of the best terrestrial photographs of the region visited by Ranger 9 in March 1965, including the crater Alphonsus (centre). Compare with

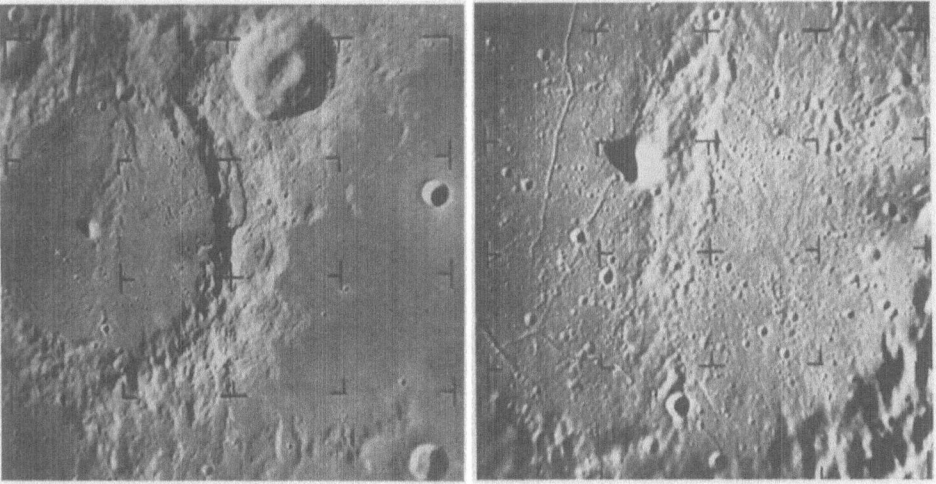

Fig. 1.7. Two views of the crater Alphonsus (cf. Figure 1.6) from space vantage points in closer proximity of the lunar surface, televised by Ranger 9 on 24 March 1965. Left – the view from an altitude of 413 km above the lunar surface, 2 m 50 s before impact, recording a square field 180 km across. Right – a view recorded at an altitude of 166 km, 1 min 9.5 s before impact (by courtesy of NASA-JPL).

it the views of this crater televised by Ranger 9 from increasing proximity to the lunar surface, as shown on Figure 1.7. Another important result disclosed by the last few frames transmitted by Rangers 7–9 was a realization of the similarity in surface texture of their respective landing places in the lunar flatlands (whose selenographic coordinates are listed in Table 1-2). Although these are separated from each other by several hundred kilometres of linear distance – two on the mare ground, and the third on the floor of a large crater – the similarity in their surface structure on the metre-scale was indeed striking (see Figure 1.8) – a fact which suggested that the processes responsible for it were of global rather than local nature.

Although the TV transmissions by Rangers 7–9 constituted no doubt the most striking contribution of the hard-landers to lunar studies in 1964–1965, it is impossible to forego without adequate mention at least some of the achievments of the preceding hard-landers – such as the measurements of the lunar magnetic field by a magnetometer contained in the 390-kg capsule of Luna 2 in September 1959, or of the γ-radioactivity of the lunar globe by Rangers 3 and 5 in 1962. The magnetometric measurements carried out aboard Luna 2 in the last minutes of its flight (cf. Dolginov *et al.*, 1960) failed to detect indications of any magnetic field exceeding in strength some 3×10^{-4} G – a fact which disclosed that the mean magnetization of the lunar

TABLE 1-2

Place and time of unmanned spacecraft landing on the Moon*

Spacecraft	Place of impact		Date and time of impact**
	Longitude	Latitude	
		(a) *Hard-landers*	
Luna 2	0°.0	29°.1 N	1959 September 13	22^{h}02^{m}24^s [†]
Ranger 6	21.52 E	9.33 N	1964 February 2	9 24 33.1
Ranger 7	20.58 W	10.63 S	1964 July 31	13 25 49
Ranger 8	24.65 E	2.67 N	1965 February 20	9 57 36.8
Ranger 9	2.37 W	12.83 S	1965 March 24	14 08 20
		(b) *Soft-landers*	
Luna 9	64°.37 W	7°.08 N	1966 February 3	18^{h}44^{m}52^s
Surveyor 1	43.21 W	2.45 S	1966 June 2	6 17 37
Luna 13	62.05 W	18.87 N	1966 December 24	18 01
Surveyor 3	23.34 W	2.94 S	1967 April 20	0 04 53
Surveyor 5	23.18 E	1.41 N	1967 September 11	0 46 44.3
Surveyor 6	1.37 W	0.46 N	1967 November 10	1 01 5.5
Surveyor 7	11.41 W	41.01 S	1968 January 10	1 05 30
Luna 17	35.0 W	38.28 N	1970 November 17	3 47
Luna 21	28.5 E	26.1 N	1973 January 15	23 35

* Only those spacecraft are included which furnished lunar scientific information.
** Universal Time as observed on the Earth (not corrected for transit time of the signals).
† The last stage of the carrier rocket of Luna 2 (1121 kg in weight) impacted on the Moon 30 min later.

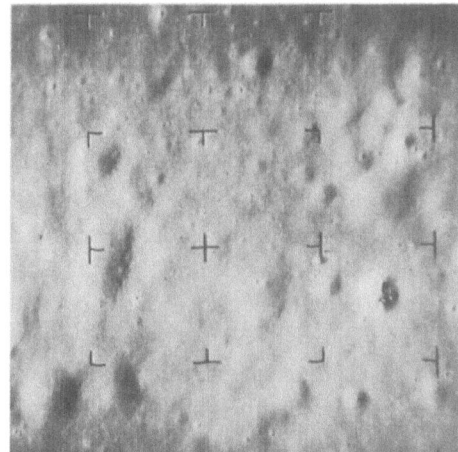
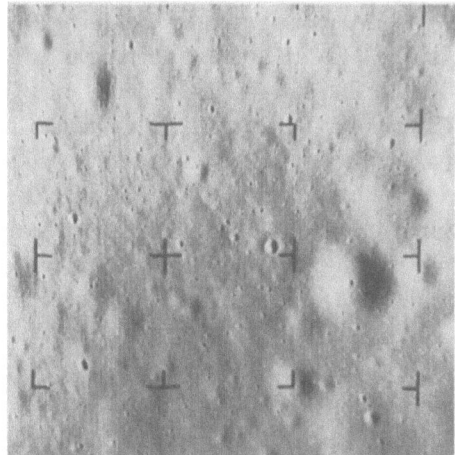

Fig. 1.8. Close-ups of the landing places of Ranger 7 (upper left; $\lambda = -20.7$, $\beta = -10.7$; on July 31, 1964); of Ranger 8 (upper right; $\lambda = 24.7$, $\beta = 2.7$ on February 20, 1965); and of Ranger 9 (lower left; $\lambda = -2.4$, $\beta = -12.9$, on March 24, 1965) – photographed from altitudes 4–6 km above the lunar surface, seconds before the impacts. The size of the individual fields is approximately $2\frac{1}{2}$ km across; and the smallest details resolved on them are of the order of one meter for Rangers 7 and 8, and substantially smaller for Ranger 9.

The reader may note that, on this resolution, the structure of the surface of all three different regions appear to be essentially the same – a fact suggesting that the forces which shaped it up are not local in nature (NASA-JPL photographs).

globe cannot exceed some 0.0025 of that of the Earth. This finding was later confirmed and further refined by the Explorer 35 (Sonett *et al.*, 1967) and Apollo data more fully discussed in Chapter 5. Similarly, measurements of γ-ray emission by Rangers 3 and 5 revealed that the mean radioactivity of the lunar globe was no greater than that of the Earth's crust – a result confirmed by the Russian soft-lander Luna 13 in December 1966 and later by the Apollo 15 and 16 missions (cf. Arnold *et al.*, 1972) as well.

3. Soft-Landing Spacecraft

With the impact of Ranger 9 in crater Alphonsus on 24 March 1965 the contributions of the hard-landers came to an abrupt end; and the next two years brought to the fore two other families of mooncraft which dominated lunar research in 1966–1968: namely, the soft-landers and the orbiters. The former preceded the latter to the Moon by some months – though not without mishaps. Five attempts at soft landings (Lunas 4–8) were made by the Russians before the success of Luna 9 on 3 February

1966 – a success repeated by Luna 13 shortly before the end of the same year (cf. Table 1-3 for the place and time of the respective soft-landings). On the American side, the first soft-lander launched on 1 June 1966 – Surveyor 1 – was fully successful. The second (Surveyor 2 in November of the same year) failed through equipment malfunctioning in the course of the mid-course manoeuvre; and it was not till Surveyor 3 (launched in April 1967) that the good fortunes of the project were recouped. Surveyor 4 (July 1967) failed again through loss of communications only $2\frac{1}{2}$ min before landing; but the fifth soft-landed safely (after some setbacks en route) in September 1967; the sixth, in November; and the seventh (and last one) in January 1968.

Before we outline an account of this accomplishments of this novel type of space-craft, let us describe them briefly and mention some of their equipment. By physical size the Surveyors (see Figure 1.9) were much larger than the soft-landing Lunas (Figures 1.10 and 1.11): 3.7 m across and 3.0 m in height, in comparison with the 1.6×1.1 m size of the Russian mooncraft; though by weight the latter were much the heavier of the two – weighing between 1422–1580 kg (see the last column of Table 1-1) in comparison with 990–1040 kg of the Surveyors, and more than four times as much as the Rangers. The actual weight of the capsules or parts which soft-landed on the lunar surface were, however, in almost inverse proportion (around 100 kg for the Lunas and 300 kg for the Surveyors); the difference between the initial and terminal mass being mainly due to propellants and attitude-control gas used up in flight, and the jettisoned main retro jet engine with altitude-marking radar. It also reflects a difference in the source of power employed by both types of spacecraft. The Surveyors were powered by solar panels, whose silicon cells provided (during day-time) electrical power of 85 W; only a small part of which (about 10 W) was used for transmission to Earth at a frequency of 2295 Mc s^{-1}. The Lunas, on the other hand, were powered by chemical batteries (and transmitted at a lower frequency of 183.530 Mc s^{-1}), giving them an active lifetime of only 3–5 days (while some of the Surveyors functioned for almost as many months).

The flight plans of the Surveyors and Lunas also differed considerably; for while the Russian spacecraft travelled to the Moon in their 80-hr trajectories, the Surveyors accomplished their journeys in about 63 hr. For the terminal descent of Surveyor 1, ignition of the solid-fuel main retro-engine was initiated by a radar altimeter at an altitude of approximately 75 km above the surface. After the main engine burned out at a height of 8.5 km (where the remaining velocity was reduced to 130 m s^{-1}), it was jettisoned; and subsequent deceleration was the task of vernier engines controlled by an autopilot and an on-board computer that used radar measurements of altitude and velocity in a closed-loop system. To reduce disturbance of the landing area by engine exhaust, the vernier engines were turned off when the spacecraft was 3.4 m above the lunar surface, and was descending at a speed of 1.5 m s^{-1}. The spacecraft then fell freely to the surface, and its vertical velocity at touchdown (which occurred on 2 June 1966, at 6 h 17 min and 37 s UT) was $3.6+0.1$ m s^{-1} (the horizontal component being less than 0.3 m s^{-1}).

The principal scientific apparatus aboard the early soft-landers were their tele-

TABLE 1-3

Kinematic characteristics of the artificial lunar satellites (1966–1972)

Spacecraft	Period	Inclination*	Altitude (in km)**		Injection into lunar orbit	End of mission	No. of days in orbit	Total no. of revolutions
			Periselenium	Aposelenium				
Luna 10	178^{m}3^s	71°9	349	1017	1966 Apr 3	1966 May 30†	67	
Orbiter 1	208.6	12.0	56	1853	1966 Aug 14	1966 Oct 29	76	547
Luna 11	178	27	159	1200	1966 Aug 18	1966 Oct 1†	34	
Luna 12	205	4	105	1740	1966 Oct 26			
Orbiter 2	208.4	11.9	49.7	1853	1966 Nov 10	1967 Oct 11	335	2289
Orbiter 3	208.6	20.9	54.9	1847	1967 Feb 8	1967 Oct 9	243	1843
Orbiter 4	721	85.5	2706	6114	1967 May 8	1967 Oct 6	70	225
Explorer 35	684	11.2	763	7670	1967 July 19	in orbit		
Orbiter 5	510.5	85.0	195	6029	1967 Aug 5		179	
Orbiter 5	503.5	84.6	100	6066	1967 Aug 5		179	
Orbiter 5	191.3	84.8	99	1499	1967 Aug 5	1968 Jan 29	179	1201
Luna 14	160	42.0	159	871	1968 Apr 10			
Luna 15	150	126	132	287	1969 July 16	1969 July 21	5	52
Luna 18	119	35	96	101	1971 Sept 7	1971 Sept 11	5	54
Luna 19	121.8	40.6	139		1971 Oct 3	1972 Sept 30	368	4350

* Inclination to lunar equator.
** Altitude above the mean lunar sphere of radius 1738 km.
† Date of loss of radio-contact.

Fig. 1.9.   The U.S. Surveyor spacecraft configuration in the terrestrial laboratory – a prototype of five which effected soft landings on the Moon between 1966–1968. The periscopic device, which served as the spacecraft's 'eyes' and helped to televise to the Earth many photographs reproduced in this book, can be seen to the left of the main mast carrying the solar panels (NASA-JPL photograph).

vision cameras, which disclosed for us the first views of the lunar surface at a really close range (attaining, in the immediate proximity of the spacecraft, resolution of the order of 1 mm on lunar ground). Little is known of the respective equipment of the Russian Lunas; but the optics and the TV-system of the Surveyors were similar to those employed previously by the Rangers. The image-forming device consisted of a dioptric system of 1-in. free aperture and variable focal length ranging from 25 mm (field of view $25°4 \times 25°4$) to 100 mm (field $6°4 \times 6°4$); with exposure times of the order of one hundredth of a second. Periscopic arrangement made it possible for the Surveyors to scan the surrounding landscape from a minimum distance close to 120 cm to the horizon about a mile away. As to Luna 9 or 13, the fact that their optics was only 58 cm above the ground gave them a much smaller maximum range of vision, but somewhat higher ground resolution in the immediate neighbourhood of the spacecraft.

Luna 9 landed in the midst of the plains of Oceanus Procellarum, near the crater Cavalerius (for the coordinates of the touchdown points cf. Table 1-2). Surveyor 1

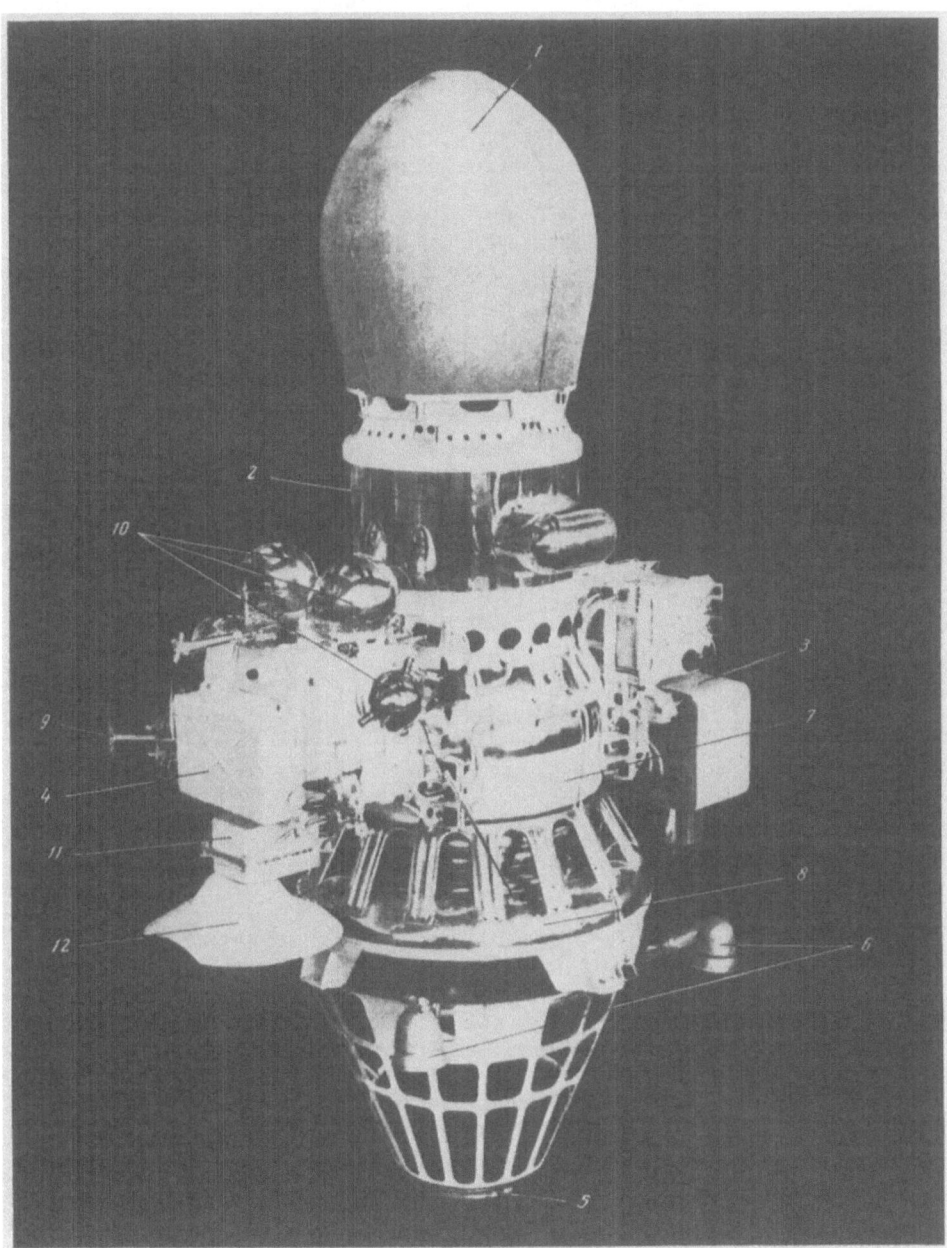

Fig. 1.10. A model of the U.S.S.R. spacecraft Luna 9, which effected the first soft landing on the lunar surface on 4 February 1966. The numbers of the individual parts marked on the photograph refer to: (1) soft-landing part of the unit (for its fuller view cf. Figure 1.11); (2) guidance system compartment; (3) camera system; (4) electronic part of the attitude control; (5) liquid fuel retro-rocket engine; (6) attitude control jets; (7) spherical fuel tank; (8) oxidizer; (9) vernier jet engines; (10) gas supply for attitude control jets; (11) radar altimeter; and (12) directional parabolic antenna (U.S.S.R. Acad. Sci. photograph).

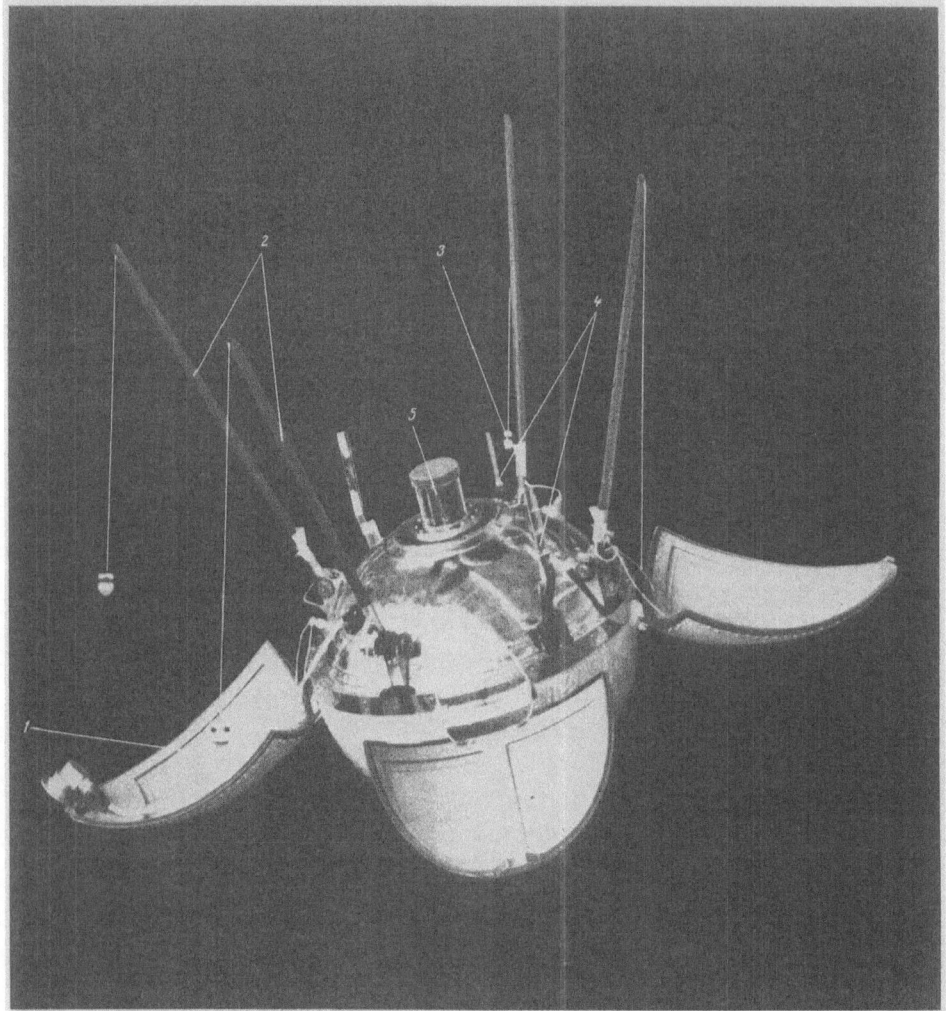

Fig. 1.11.   A model of the 100-kg capsule of Luna 9 with antennae and optics unfolded. The spherical part
of the capsule is approximately 50 cm in diameter (U.S.S.R. Acad. Sci. photograph).

touched ground in a nearby region in the neighbourhood of the crater Flamsteed in
the early hours of 2 June 1966 – two days after local sunrise – and operated through
the entire lunar day that ended on 14 June; and during this time approximately
11 150 pictures were obtained in daylight at Sun elevations ranging from 2° to 88°
(see Figures 7.6 and 7.7). After sunset, pictures of the spacecraft were taken in earthshine; in addition to some photographs of the solar corona of stars and of the planet
Jupiter. When the temperature dropped well below freezing, the spacecraft was shut
down after 53 hr of lunar night operations. It was turned on again on 6 July and
operated until the next sunset on 14 July. By that time its primary mission was completed; but the system was operated again between 8–10 October; and subsequent

engineering interrogation of the spacecraft was conducted from time to time over a period of the next eight months.

Surveyor 2, launched on 10 September, had essentially the same configuration as its predecessor and was intended to land in Sinus Medii. During the mid-course manoeuvre on 21 September one of its three vernier rocket engines failed, however, to ignite and the unbalanced thrust caused the spacecraft to tumble. Repeated commands were of no avail; and Surveyor 2 – out of control – crashed to its destruction on 23 September to contribute an addition to our tabulation of the hard-landers.

Surveyor 3, launched on 17 April 1967, fared better; for it landed softly close to the appointed place in the south-east portion of Oceanus Procellarum (about 370 km south of the crater Copernicus), and came to rest on the inner slope of a rounded shallow crater about 200 m in diameter (see Figures 1.12 and 1.13). Its touchdown was, however, rather dramatic, because the vernier engines did not shut down at the planned altitude of 4.2 m. As a result, at the first touchdown the spacecraft had a lateral velocity of about 1 m s^{-1}, which caused recoil and two additional touchdowns to occur at a distance of 30 and 11 m from the initial landing place (cf. Figure 1.14).

As Surveyor 3 landed inside a shallow crater – a fate which previously befell also the Russian soft-lander Luna 13 – its television cameras could not see outside the

Fig. 1.12.   Surveyor 3 on the Moon – a photograph of the spacecraft in its lunar environment taken by the astronauts of the Apollo 12 mission in November 1969 (NASA official photograph).

Fig. 1.13.   Rendezvous of men and machines on the Moon – Surveyor 3 (foreground) and Apollo 12 lander ('Intrepid') on the horizon are shown on this photograph taken by astronauts Charles Conrad and Alan Bean during their second walk on the Moon on 29 November 1969 (NASA official photograph).

latter's ramparts. However, between 20 April and 3 May 1967, the spacecraft relayed to us 6315 individual televised pictures, some of which are reproduced on Figures 1.13 and 1.14. Just as Surveyor 1 – not being content to scan only the lunar panorama – turned its lenses to the solar corona and to the stars – Surveyor 3 happened to be the first soft-lander to witness, on 24 April 1967, an eclipse of the Sun by the Earth on the lunar surface, when for 41 min of totality sunlight was completely cut off. During the entire eclipse lasting 107 min, Surveyor 3 experienced, and reported to Earth, the precipitous temperature changes amounting to almost 200° centigrade; thus confirming the results of previous work performed from our terrestrial home bases. However, its cameras were the first eyes to relay to us the view from space of the terrestrial aureola surrounding the rim of our planet (due to a diffraction of sunlight in our atmosphere) and also of the entire Earth in colour.

Fig. 1.14. An imprint of one of the footpads of Surveyor 3 in the lunar soil caused by recoil at landing, and photographed *in situ* three years later by astronauts of the Apollo 12 mission (NASA official photograph).

Surveyor 4, launched in July 1967, proved once more (but for the last time) a technical failure, because of a last-minute breakdown of its communication system; and Surveyor 5 gave also rise to some concern en route about its performance. Fortunately, the incipient trouble was rectified in time; and the spacecraft landed faultlessly on September 11 near the eastern shores of Mare Tranquillitatis to embark on a highly successful mission which until the second lunar sunset on 23 October furnished 19054 new television pictures (of which over 18 thousand were obtained during the first lunar day between 11–23 September). These included views of the interior of the crater in which the spacecraft landed, and of the level mare ground surrounding the crater; of star and planet sightings for attitude reference; as well as a sequence of solar corona and earthshine pictures after local sunset on the spacecraft. These provided us with the first measurements of the brightness of the solar corona between 10–30 solar radii from the centre of the Sun (i.e., up to a distance of one-third of the dimensions of the orbit of Mercury). In addition, Surveyor 5 witnessed on the lunar surface another eclipse of the Sun by the Earth on October 18, and augmented thermal measurements of precipitous changes in surface temperature during this time.

The configurations of Surveyors 3 and 4 were basically similar to their first two predecessors; but carried several new items. Like the previous spacecraft of this class, they were equipped with television systems and instrumentation for determining the bearing strength of the lunar surface in their footpads. In addition, however, Sur-

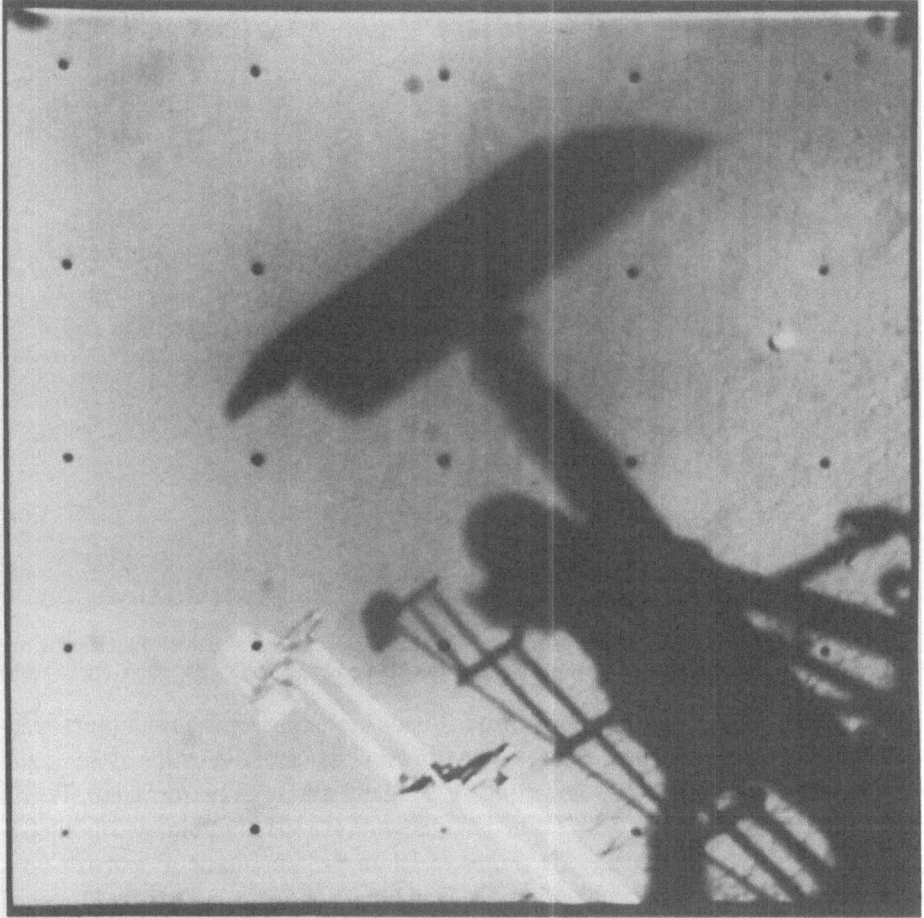

Fig. 1.15.   The outlines of the mechanical surface sampler of Surveyor 3 appear on a photograph televised by this spacecraft on 2 May 1967, together with the shadow cast by it on the lunar surface by the rays of the late afternoon Sun. The arm of this sampler could reach up to 1.5 m distance from the spacecraft. Between the sampler and the main mast, a shadow of the TV-camera which took this picture is readily apparent (NASA-JPL photograph).

veyor 3 and its successors carried to the Moon a remotely-controlled mechanical surface sampler (Figure 1.15) – a device for digging and otherwise manipulating the lunar surface material in full view of the television camera (see Figure 1.16). Lastly, Surveyors 5–7 were provided with a chemical surface sampler (cf. Figure 1.17) to analyze by remote control the atomic composition of the outermost lunar crust in the following manner. Six artificial radioactive sources of curium Cm 242 (plus einsteinium Es 254 whose 6.42 MeV α-particles served as a standard) for calibration are contained in a sensor head – a box approximately $17 \times 15 \times 13$ cm in size; with a 30.5 cm bottom plate (to prevent the sinking of the box into the possibly soft lunar surface) – which contains also two α-particle detectors which can measure the energy

Fig. 1.16.   The surface sampler of Surveyor 7 in action – the photograph (televised on 20 January 1968) shows a rock fragment and a small amount of loose material scooped up from the lunar surface – some of it adhering to the smapler's scoop door (NASA-JPL photograph).

Fig. 1.17.   Surveyor's 5 α-particle scattering instrument, used for chemical analysis of the lunar soil on the last three missions of the project (Surveyors 5–7), with the results described in Chapter 7. (NASA-JPL photograph).

of α-particles back-scattered from the lunar surface, together with four proton de-tectors measuring the energy of protons produced by (α, p) reactions with the atoms of the lunar surface. When the sensor rests on the lunar surface, the energy spectra registered by these detectors and telemetred to the Earth (with an averaging time of the order of 20 hr), can then be analyzed to indicate atomic proportions of different elements constituting the surface; while when the sensor is automatically lifted to a height of 38 cm above the surface (at which back-scattered α-particles can no longer reach the detectors) the sensor registers the natural radioactivity of lunar rocks.

The α-scattering instrument aboard Surveyor 5 transmitted data on the chemical composition of the lunar surface for a total of 82 hr during its first lunar day on the Moon (between 11–29 September 1967); and additional 22 hr up to the next sunset on 23 October. The results of this analysis (together with those based on the rocks brought back by the Apollo missions) will be given in Chapter 7. In addition, a magnet assembly measured the percentage component of surface material of high magnetic susceptibility, and found it to be comparable with that of pulverized basalt containing 10.12% of magnetite, and less than 1% of metallic iron. Moreover, particles attracted to the magnet were mostly less than 1 mm in size. In order to provide further infor-mation on the effects of engine exhaust on lunar surface material, the liquid-pro-pellant vernier engines were fired on September 13 for 0.55 s; measurements made also of local radar reflectivity; and surface temperature data obtained during the first lunar day and after sunset until September 29.

Surveyor 6, which landed on the Moon on 10 November in a nearly flat, heavily cratered ground of the lunar Sinus Medii with a touchdown speed of 3.4 m s^{-1}, was from the instrumental point of view essentially a replica of Surveyor 5. It carried a similar television system and α-particle back-scattering experiment to prove the chemical composition of the underlying ground. However, the colour filters used on the Surveyor 5 camera were replaced by three polarizing filters.

On November 17, the vernier rocket engines of Surveyor 6 were fired again for 2.5 s to lift the spacecraft some 3.5 m – thereby achieving the first rocket-powered takeoff from the lunar surface – and it landed again 6.1 s later in another place 2.4 m away. This hop permitted a stereoscopic view and mapping of the surrounding lunar features, provided clearer view of surface depressions produced by the initial landing of the spacecraft, and furnished additional information on the effects of the burning jet engines on the lunar surface.

Between November 10 and 24, Surveyor 6 transmitted 30 396 individual television frames featuring the views of the undisturbed lunar surface as well as of the areas disturbed by the initial landing and the hop; views of the spacecraft (including the magnet assembly attached to one footpad); views of the Earth, of stars, as well as of the solar corona. For the first time, the degree of polarization of the corona was measured up to an angular distance of 15° away from the Sun. The α-particle instrument transmitted for 30 hr new data on the chemical composition of the outermost crust. Surface temperature data were obtained throughout the lunar day and for 41 hr after sunset. The magnet assembly provided further information on material of high magnetic susceptibility. Radar reflectivity data were obtained during terminal descent; and the measurements of loads in the spacecraft legs during the initial touchdown (as well as during the hop) provided additional data on the mechanical properties of lunar surface material.

Surveyor 7 – the last member of the family of American soft-landers – reached its destination on 9 January 1968, in a mountainous region adjacent to the crater Tycho. In the course of its first day on the Moon it performed chemical analysis of the rocky ground on which it landed (see Figure 1.18) and its mechanical scoop dug up several tranches to test the hardness of the ground. By January 21, Surveyor 7 returned about 17 000 new television pictures (of which Figures 1.18 or 7.6–7.7 can be considered as examples) – thus bringing the grand total for the five successful craft of its class to over 83 000.

A few days before, Surveyor 7 performed another unique experiment: namely, by tilting its periscopic mirror upwards to view the Earth, it recorded the flashes of laser beams aimed at its landing place near Tycho from several geographic localities in the western United States. Televised images returned to the Earth disclose that the beams were indeed detected from the spacecraft; and although the quality of signals was rather marginal, this experiment demonstrated laser beams of moderate intensity can be used for communications with astronauts on the Moon should radio link fail through damage to the receivers.

In contrast with the American soft-landers, on which the design of experiments as well as their preliminary results have been fully published, much less is known on the instrumentation of the soft-landing Lunas – and still less about their results. It has, however, been disclosed that Luna 13 (which landed on the Moon on 23 December 1966 in Oceanus Procellarum near the crater Seleucus) contained in its

Fig. 1.18.   Horizontal panorama north-west of the landing place of Surveyor 7, about 29 km north of the crater Tycho (see Figure 1.31). The horizon formed by a series of hills and ridges is about 13 km away.

capsule a soil density-metre (hammer head with titanium point), driven into the lunar soil by a jet device, mounted on the end of one experimental boom. A second density gauge – γ-ray source to irradiate the surface and gas-discharge counters to measure the flux of dispersed radiation – were mounted on the other boom; but the results of the experiments performed with the aid of these devices have not so far been published.

What kind of a landscape did the TV-cameras of the American and Russian soft-landers reveal to us on Earth in the course of the horizontal sweeps of their peri-scopes? All but the last of them landed on what appears to be typical mare ground, which proved to be a gently rolling and relatively smooth surface – nearly level on a kilometric scale – sparsely dotted with boulders of various sizes and shapes; and covered with small craters (from several centimetres to several metres in dimension) which are of secondary (or tertiary) impact origin. Surveyor 7 alone landed on a typically continental ground, in the proximity of the crater Tycho; but even there the horizontal panorama (shown on Figure 1.18) does not appear to be very different from that in the maria. In the immediate neighbourhood of the spacecraft, the lunar material was seen (cf. Figures 1.19 or 1.20) to consist of granular material ranging widely in size – coarser blocks of rock of smaller fragments interspersed with particles less than a millimetre in size – which was penetrated by the spacecraft's footpads down to a depth of a few centimetres (Figure 1.21).

With the exception of the one 'hop' performed by Surveyor 6 on 17 November 1967, none of the spacecraft soft-landed on the Moon between 1966–1968 was capable

Fit. 1.19. A composite picture of several individual frames televised by Surveyor 1 on 13 June 1966, which shows the immediate surroundings of one of the landing pads of this spacecraft, and the nature of the landing ground (NASA-JPL photograph).

of locomotion. It is true that the possibility of sending a roving vehicle to the Moon was considered briefly by NASA in the early 1960's by a short-lived project under the code name of 'Prospector', but it was dropped soon thereafter not to compete with the requirements of the Surveyors. A roving vehicle did not, in fact, reach the Moon until November 1970 in the form of the Russian 'Lunokhod' (Luna 17) – whose model is shown on the accompanying Figure 1.22 and was followed by Luna 21 in January 1973.

The first of these unmanned Soviet spacecraft was soft-landed on 17 November 1970 at 3h 47 min UT in the northwestern part of Mare Imbrium, some 80 km south of Promontorium Heraclides (for the coordinates of a more exact position cf. Table 1-2). It represents a remotely-controlled eight-wheel electric car powered by solar

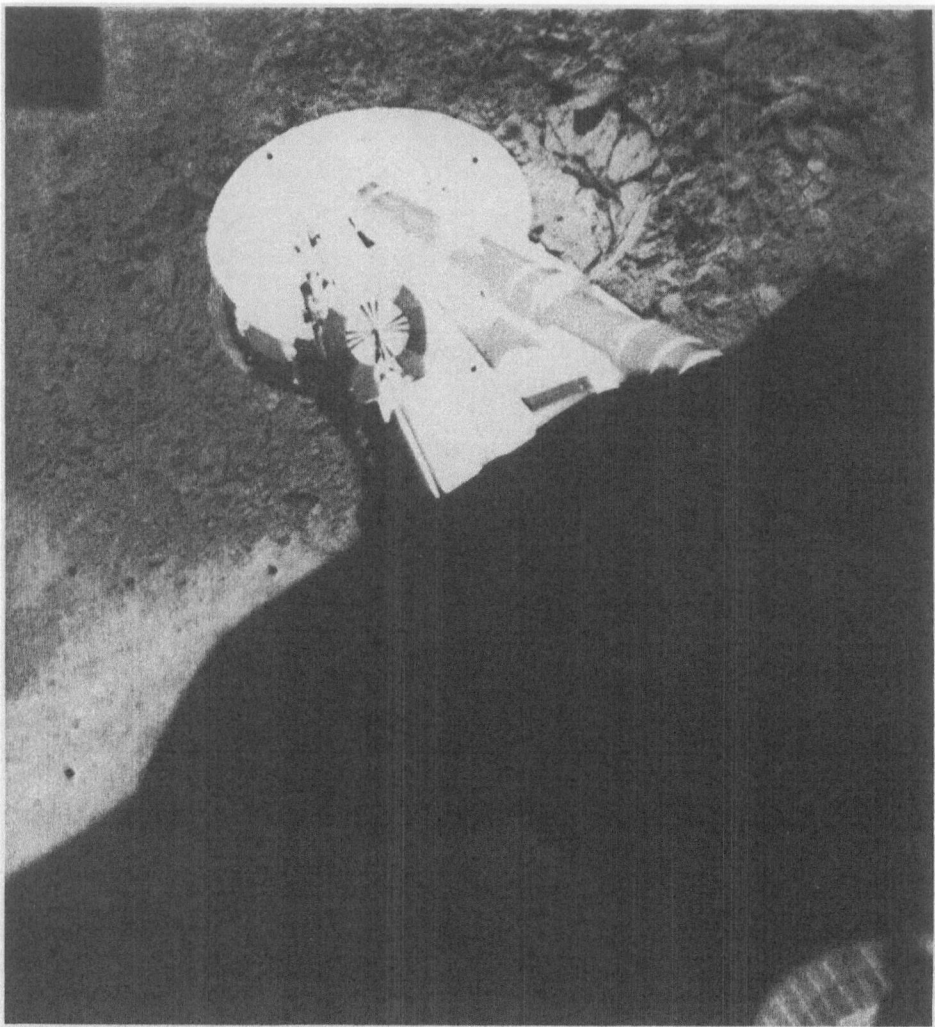

Fig. 1.20.   Surveyor 6 footpad resting on the broken lunar ground after its 'hop' (NASA-JPL photograph).

batteries and designed to carry scientific equipment and television cameras over the lunar surface. Its wheels (see again Figure 1.22) are approximately 50 cm in diameter, and the total weight of the moving vehicle is approximately one ton. Each wheel is equipped with an electric motor and two-speed transmission; with all wheels capable of forward and backward motion. During its first few days the vehicle negotiated slopes as steep as 14°.

The 'eyes' of the rover are provided by an electronically-scanned TV system, whose wide-angle and high-power lenses permit panoramic views of the landscape as well as close-up looks at objects of interest (cf. Figures 1.23 and 7.8). While not many details of the experiments carried out by the Lunokhod on the Moon are available so far, it is known that one of its most important tasks has been the collection and

Fig. 1.21.   A close-up view of one of the footpads of the Surveyor 1, which sunk on touchdown (4 June 1966) several cm into the lunar ground – a penetration indicative of the surface bearing strength of a few pounds per square inch (NASA-JPL photograph).

analysis of soil samples by X-ray fluorescence technique. A proton counter was employed to study the cosmic rays, and an X-ray telescope to observe the galactic X-ray sources from a more stable platform than rockets or satellites provide. On top of the Lunokhod was an array of 14 cube-corner reflectors for precise measurements of the Earth-Moon distance by means of laser echoes – a programme carried out in collaboration with the French National Council for Space Research. Communications with the Earth have been maintained by means of two types of antennae: one being a low-gain, broad-beam cone; while the other is a directional high-gain helix which is gimbal-mounted so that it can be aimed towards the Earth regardless of Lunokhod's orientation.

In contrast with the relatively limited contributions to lunar science of the hard-landers, whose scientific packages were rather severely monothematic, the contributions of the soft-landers were profound; and the reader will find them interspresed in all subsequent chapters of this book. While the early fly-by spacecraft of Russian origin gave us the first inkling of the far side of the Moon, and the hard-landing Rangers resolved (over the coverage area of their last frames) details less than a metre

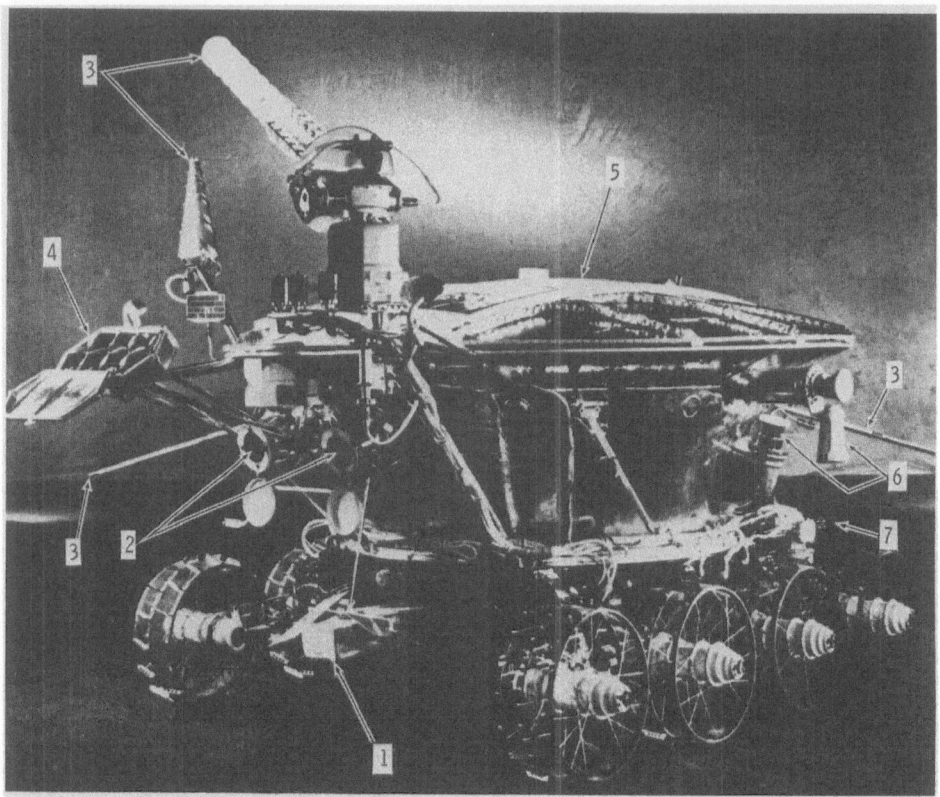

Fig. 1.22.   A view of the model of the U.S.S.R. Lunokhod surface vehicle, landed and operated on the Moon in November 1970 (Luna 17) and January 1973 (Luna 21). The individual numbers on the model indicate the locations of (1) the X-ray spectrometer; (2) television cameras; (3) antennae; (4) cube-corner reflectors for the return of laser signals; (5) lid over the cooling radiator for the solar cells supplying electrical power, which are on the inside of the lid (the vehicle is shown in night-time configuration, when the lid is closed); (6) duplicate cameras); and (7) soil mechanics instrument and odometer wheel (after Alexandrov *et al.*, in *Space Research* XII COSPAR, Berlin 1971).

in size – thus exceeding telescopic resolution attainable from the distance of the Earth by a factor of the order of 1000 – the soft-landers increased this spatial resolution in the immediate neighbourhood of the mooncraft by a further factor of 1000; and performed many other experiments which would previously have been unthinkable. Thus one of the most important contributions of the Surveyors has been to provide the first evidence that the chemical composition of lunar rocks in the continental area in the proximity of the crater Tycho is different from that of the typical mare ground visited by the preceding Surveyors or the Lunas. This disclosed for the first time that the surface of the Moon is chemically differentiated – a subject to which we shall return more fully in Chapter 7.

4. Orbiting Spacecraft

The second signal achievement in space research of the Moon between 1966–1968

Fig. 1.23. Lunokhod 17 on the Moon – a televised close-up of the rocky surface of Mare Imbrium, showing tracks left behind by the moving vehicle. The instrument protruding in the field of view on the right is the shield of a camera marked by '6' on Figure 1.22 (U.S.S.R. Acad. Sci. photograph).

has been a successful launch and injection into circumlunar orbits of a family of no less than 11 artificial satellites – five by Russia and six by the United States – one of which still continues to hold its lonely vigil around the Moon. Some of their characteristics have already been listed in Table 1-1(d); and others are contained in the accompanying Table 1-3. Before, however, we review their principal accomplishments, a few words describing the equipment which they carried aboard may again be appropriate.

The U.S. Orbiter spacecraft (see Figure 1.24) were distinctly smaller and lighter than the Surveyors – only 2.08 m in height; 5.21 m in span between the tips of the high- and low-gain antennae; and 3.79 m across the solar panels. When fully loaded with film and fuel for retro-engines (plus gas for the requisite manoeuvres) their weight was close to 380 kg. Their energy supply was provided mainly by the solar panels, consisting of over ten thousand individual silicon cells delivering electrical energy close to 375 W; in addition, a 20-cell nickel-cadmium battery aboard provided a steady current of 22–31 V for operations to be performed in the Moon's shadow. Three-axis stabilization was provided by using the Sun and Canopus as primary reference points; and by a three-axis inertial system in use when both the Sun and Canopus were occulted by the Moon.

The reason for the smaller size and weight of the Orbiters (comparable indeed with the Rangers rather than the Surveyors) was the fact that – unlike the multi-experiment Surveyors – the Orbiters constituted an almost monothematic task-force: namely, one aiming to secure from close proximity a photographic coverage of the globe of the Moon at moderate and high resolution. The optical system (see Figure 1.25) employed to this end consisted of two photographic lenses producing their respective images side by side on 35 mm strips of a 70-mm Eastman-Kodak SO-243 special high-definition aerial film – not overly sensitive, perhaps, but one with extremely fine grain (capable of resolving 250 lines per mm). High-resolution photographs were obtained with a Pacific Optical Paxoramic lens of 11 cm aperture and imaging a field of 61 cm focal length; while moderate resolution was attained with a Schneider Xenotar lens of 1.4 mm aperture and 80 mm focus, imaging afield over 20° across. Both lenses operated, therefore, at a focal ratio of $f/5.6$, with controllable shutter speeds of 10, 20 and 40 ms.

These exposures were sufficiently short to 'stop' the motion of the lunar target from the 'working' altitudes without noticeable loss of ground resolution in the focal plane of the small Xenotar lens; but not so with the high-resolution optics. For at a typical orbital velocity of 1936 m s^{-1}, the spacecraft moves by 77 m during 40 ms-

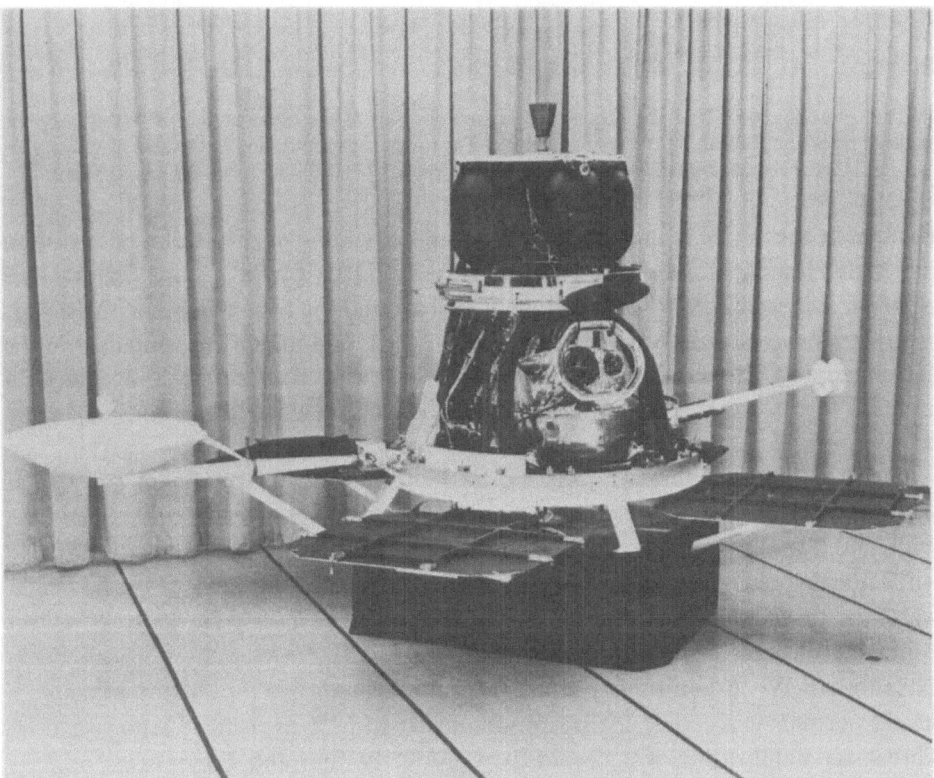

Fig. 1.24. A view of the U.S. Orbiter spacecraft in the laboratories of the Aerospace Division of the Boeing Company, with antennae and solar panels unfolded. The optics of the Orbiter cameras are exposed to view in front (NASA-Boeing photograph).

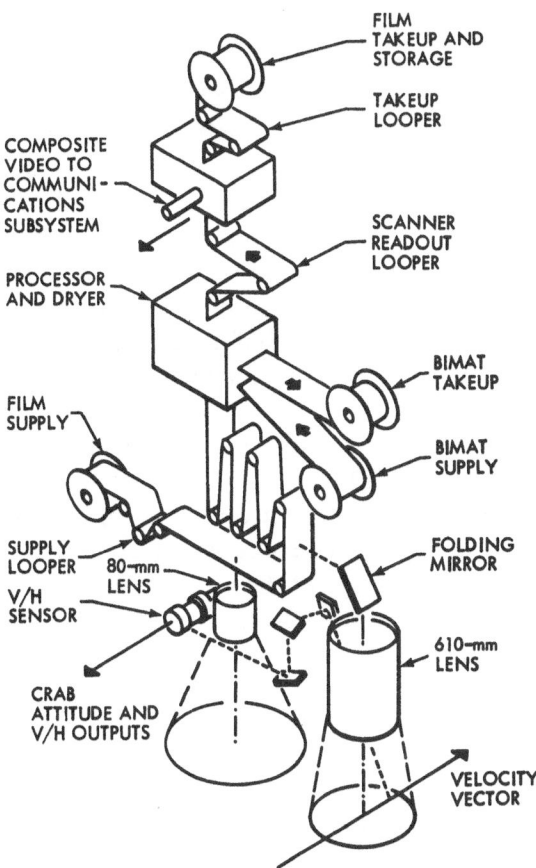

FILM
TAKEUP AND
STORAGE

TAKEUP
LOOPER

COMPOSITE
VIDEO TO
COMMUNI-
CATIONS
SUBSYSTEM

SCANNER
READOUT
LOOPER

PROCESSOR
AND DRYER

BIMAT
TAKEUP

FILM
SUPPLY

BIMAT
SUPPLY

SUPPLY
LOOPER   80-mm
          LENS

FOLDING
MIRROR

V/H
SENSOR

610-mm
LENS

CRAB
ATTITUDE AND
V/H OUTPUTS

VELOCITY
VECTOR

Fig. 1.25. A schematic view of the photographic system of the Lunar Orbiter satellites, used for on-
board photography and its subsequent readout for transmission to the Earth.

exposures; and the smear caused by this motion must, therefore, be suitably compen-
sated. In order to do so, an image-compensation motion (depending on the instantaneous velocity-to-height ratio) was imparted to the film during exposures through
the telephoto lens. It was a partial malfunctioning (eventually traced to an over-
sensitive relay) of this V/H device which degraded somewhat the quality of high-
resolution photographs taken by Orbiter 1 in August 1966.

The exposed film was then developed, fixed and dried aboard the spacecraft. The
Eastman-Kodak 'Bimat' system processed the film at a rate of 60 mm per minute;
and required 3.4 min fully to process the latent image. This image was then scanned
by a narrow light beam producing a 6.5-micron spot on the film. The spot swept a
strip 2.67 mm wide in the direction of the film – 286 sweeps per mm being made
along the 70 mm film width – and on their completion the film is advanced by 0.1 in.
(2.54 mm; affording some overlap) and the scanning process continued in the opposite direction (see again Figure 1.25).

The readout of the photographic data for a length of film equal to a single high-

and moderate-resolution pair required approximately 43 min, and was transmitted to the Earth in real time over a high-gain antenna – 36-in. parabolic reflector (cf. Figure 1.24). Two frequencies (2116.38 and 2298.33 Mc s^{-1}) were used for two-way communications; the low-gain (omnidirectional) antenna at the end of the 2-m boom served for the reception of commands from the Earth. The rate of picture transmission meant, in effect, that the Orbiters were able to transmit to us only about two full frames during each revolution of approximately 200 min (for fuller details cf. Table 1-1(d)) while the spacecraft was visible from the Earth. At the receiving stations on the Earth, the telemetered images were reconstituted in strips: each 2.67 mm strip of film aboard the spacecraft (scanned in 18 000 lines) was enlarged so that each 35 mm frame of moderate – and high-resolution photography was reproduced on a field of 9 × 14.5 in. (229 × 368 mm).

The images recorded by the Orbiter optics were, therefore, not formed – as for the Rangers or the Surveyors – in a TV-vidicon tube, but on photographic emulsion. In this respect, the American Orbiters reverted to a strategy previously employed by the Russian Luna 3 or Zond 3, which the American spacecraft was developed to a high degree of perfection. The scope of the actual mission of each Orbiter was, of course, limited by the length of the 70-mm film, which could be accommodated aboard; and weight considerations limited this length to approximately 200 ft. This was sufficient to record about 211 frames with each lens, but no more; for unlike with the TV-vidicon tubes of the Surveyors or Rangers, a photographic film cannot have its image erased for subsequent use in the course of a mission. Therefore, the total number of individual photographs obtained by all Orbiters is close to one thousand – almost twenty times less than the combined output of the Surveyor. On the other hand, the number of picture points (per unit area) which can be extracted from photographs exposed on high-definition films exceeds by orders of magnitude the information which can be read off the screens of TV-vidicon tubes. In other words, photographs on high-resolution films can contain a vastly more detailed record than a televised image; and for this reason – in spite of its non-linear response to light – the photographic process still proves superior to pure television techniques in this particular branch of space astronomy, when requirements call for the detailed study of a target rather than its preliminary reconnaisance.

Furthermore, in further comparison of the techniques and accomplishments of the landers and orbiters, the latter had the inestimable advantage of superior vantage points, from which they could see a much larger fraction of the lunar surface – the entire globe of the Moon, in fact – from their aposelenium distances; and near the time of their periselenium passages – from altitudes close to 50 km above the lunar surface – the ground resolution of their photographs taken with their telephoto objectives was close to 1–2 m. The latter was comparable with the ground resolution attained by the Rangers in the last few seconds of their flight (and, of course, much less than that of the soft-landers); but while the three Rangers 7–9 among them recorded not more than about 1 km^2 of the lunar ground on this resolution, the Orbiter telephotos have extended the area of the Moon which we now know in this detail to about 100000 km^2 on the front side of the lunar globe. Moreover, at moderate resolution of 50–200 m (depending on the instantaneous height of the spacecraft) both sides of the Moon have now been completely covered by photographic records – no part of the lunar surface is unknown to us any more. It is, in fact, probably true to say that – thanks to the Orbiters – we are now in possession of more complete topographic records of the entire lunar surface than we are for the Earth. Moreover, many of the moderate-resolution photographs were taken from vantage points sufficiently close together to afford stereoscopic overlap – an advantage not obtaining, unfortunately, for the high-resolution photographs.

To give a profile of at least the first few of these missions, the U.S. Orbiter 1 was launched from Cape Kennedy on 10 August 1966, at 19 h 26 min UT Twenty eight hours after launch, a single mid-course manoeuvre was executed (using the Sun and the Moon to establish the spacecraft's altitude); and 98 hr after launch (on August 14) the spacecraft was injected into a circumlunar orbit inclined by 12° to the lunar equator, and altitude above the surface varying between 199 and 1850 km. The actual photographic mission did not commence until August 14 – four days and 23 hr after injection into lunar orbit – and on August 21 (during the 44th orbit) the velocity control engine was ignited on command from the Earth to reduce the periselenium distance from 200 km to 56 km above the lunar surface. Film production and processing in this 'working' orbit was completed on August 30, and the readout of all frames initiated. A total of 211 dual-exposure (i.e., moderate- and high-resolution) photographs were taken, and their transmission completed on September 14. The exhausted spacecraft then continued revolving around the Moon until almost the end of the next month, when on October 29 the last command from the Earth caused it to crash-land on the lunar surface (see column 7 of Table 1-3).

Orbiter 2 followed a similar course of action. Launched on 6 November 1966, at 23 h 21 min UT; the mid-course manoeuvre was executed 44 hr later; and injection into a circumlunar orbit (or distance from the surface varying between 196 km and 1871 km from periselenium to aposelenium) occurred on November 1, 20 h 27 min UT, 92.5 hr after launch. After 23 revolutions in this orbit, the spacecraft's velocity was reduced to effect a transfer into a 'working' orbit of 49.7 km initial periselenium distance, and 1850 km distance at aposelenium. Photography commenced in this orbit on November 18 (11 days and 16 hr after launch, during the 45th revolution); and was completed on November 26 (103rd revolution); with readout terminated by December 10. The schedule of Orbiter 3 – the next member of the series successfully launched on 4 February 1967 – followed much the same pattern and need no longer be reproduced here in full, as the relevant orbital data have been compiled in Table 1-3.

The cautious approach of the Orbiters to their target – calling for almost a week for each to spend in a wide 'parking' orbit around the Moon before being lowered to the 'working' photographic orbit – had a very good reason; for these spacecraft had first to 'feel' their way in the gravitational field of the Moon (on which very little direct information was available up to that time) before being ready for the more hazardous subsequent part of the operation. Only after an in-flight analysis of accurate range-Doppler tracking of the spacecraft disclosed at least the magnitude of the principal harmonic terms in the external gravitational potential of the lunar globe, could the Orbiters be lowered with caution into their photographic 'working' orbits (the principal characteristics of which have been collected in Table 1-3), in which each spacecraft approached the lunar surface at a much closer range. The cumulative effects of lunar gravitational perturbations (arising from the departures of the lunar globe from a spherical shape, to be discussed in Chapter 4) exert slow secular changes in several elements of the orbit and eventually made most spacecraft crash on the lunar surface; though for wide orbits this may take a long time.

What were the scientific goals which the Orbiters were called upon to carry out in the lunar proximity? The most important task of their missions – at least for the first three of them – was to photograph, at both moderate and high resolution, thirteen prospective sites for the Apollo landings, in order to investigate their surface structure in greater detail than can be done from terrestrial photography. By March 1967 this had been successfully completed; and an increased fraction of time could be devoted to the missions' secondary objectives – which was to photograph lunar surface features of particular scientific interest.

This had already been done during missions 1–3 with a part of the available film. The next spacecraft of this class – Orbiter 4 – was deployed to photograph the entire visible face of the Moon from 'parking' altitudes; and Orbiter 5 did the same for the Moon's far side. Between them, the last two spacecraft have provided us with a complete photographic coverage of the entire lunar globe with a ground resolution of 50–100 m – i.e., about ten times higher than that attainable with the largest telescopes, and under best seeing conditions, from the distance of the Earth. As a result, we are now in possession of photographic data recording the topography of the entire lunar surface more completely than we know it for our own planet; but the measurement and mapping of this enormous material (containing as it does records of over 20 million crater formations) will take years of time.

Such Orbiter photographs as are reproduced on the accompanying Figures 1.26 to 1.32 bear witness to an enormous advance in our acquaintance with detailed topography of the lunar surface brought about by these means.

A veritable *tour-de-force* of the early part of the Orbiter programme has been the identification of Surveyor 1 on the lunar surface. The region where it landed was photographed at moderate resolution already by Orbiter 1 in August 1966 – less than three months after Surveyor 1's touchdown – and the spacecraft itself was eventually located by its shadow on a low-overpass photograph taken by the telephoto lens of Orbiter 3 in the spring of 1967 (see Figure 1.33). In the same manner, Orbiter 5 photographed for us the impact crater – some 16 m across – produced by the hard landing of Ranger 8 in February 1965 (cf. Figure 1.34).

Orbiter 5 – the last one of the series – crashed to its doom on the lunar surface on 29 January 1968. Before it did so, however, this spacecraft performed one more feat worthy of record in the annals of our science: namely, by proper orientation of its solar panels on command from the Earth it was made to reflect sunlight specularly

Fig. 1.26. A photograph of the southern part of the lunar far side, taken by the wide-angle lens of the
Lunar Orbiter 3 on 19 February 1967, from an altitude of approximately 1500 km above the lunar surface.
The large crater with dark floor is Tsiolkovsky (NASA-Langley photograph).

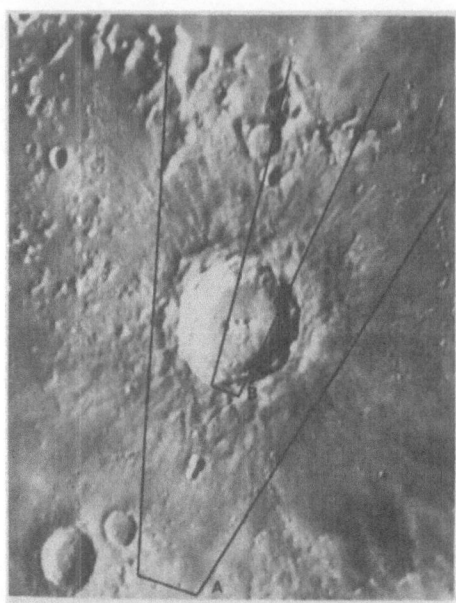

Fig. 1.27.   A view of the crater Copernicus as photographed from the Earth (upper right), in contrast with the views recorded by the Lunar Orbiter 2 on 23 November 1966 from a closer proximity to its target. Photograph on the left shows an oblique view of this crater (with camera pointed 17° below the horizon) recorded by the Orbiter's wide-angle camera; while the photograph below shows a high-resolution view of the interior of the crater. At the moment when this latter photograph was taken, the spacecraft was flying over the southern ramparts of Copernicus at an altitude of 45.4 km. The group of hills in the forefront constitute the 'central mountain' of the crater; and the broken nature of its inward-sloping walls is well seen in the background. The large hill on the horizon to the left is the 900 m high Gay Lussac of the lunar Carpathians. The fields of view of both these Orbiter photographs are marked on the terrestrial photograph (upper right) by the respective cones (NASA-Langley photographs).

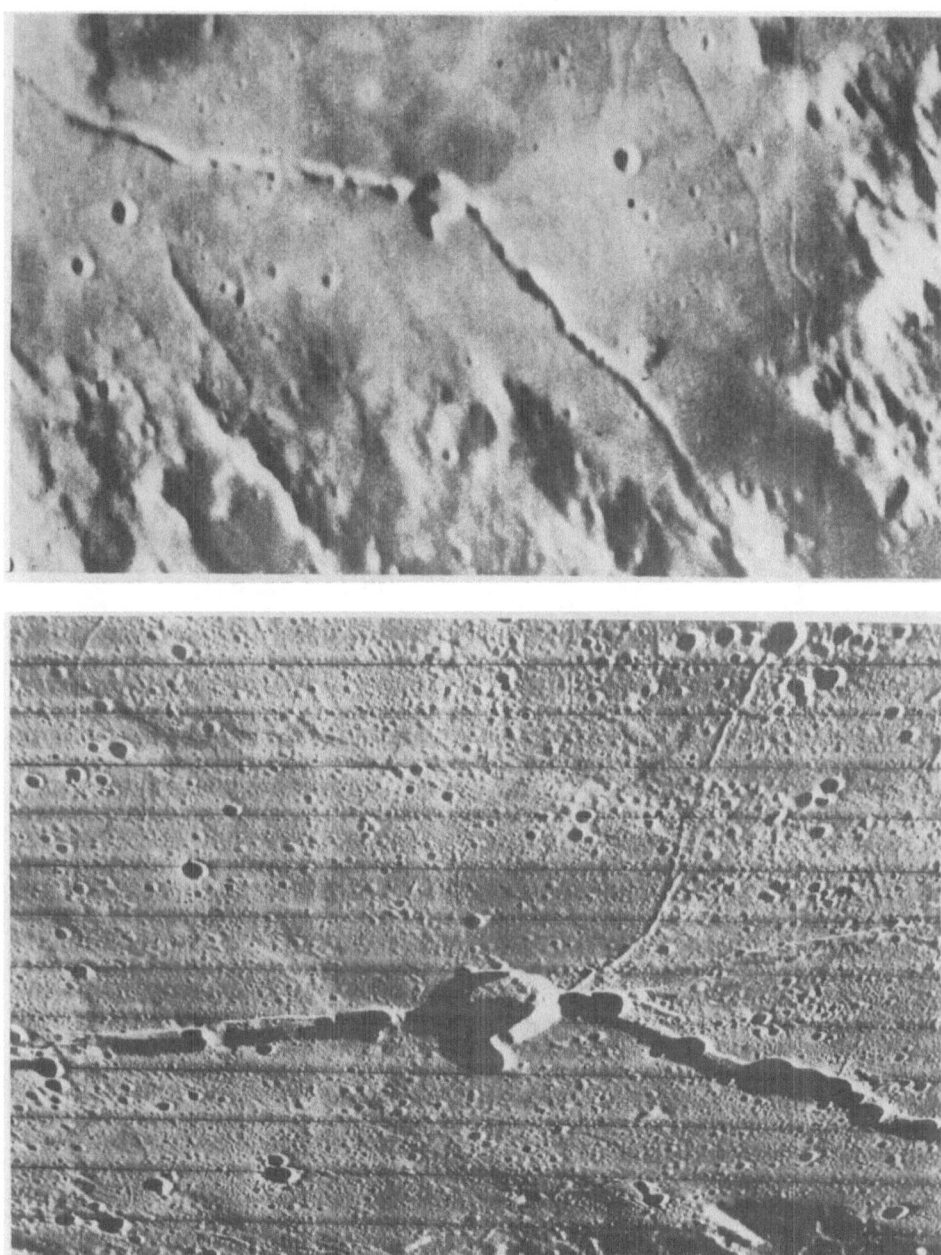

Fig. 1.28. A comparison of the Earth-based and space-borne photography of the Moon – the Hyginus rille in Mare Vaporum, as photographed by Bernard Lyot on 21 March 1945 with the 24-in. refractor of the Observatoire du Pic-du-Midi (above), and by the wide-angle lens of the Lunar Orbiter 3 on 18 February 1967 from an altitude of 62 km above the lunar surface (below) (NASA-Langley photograph).

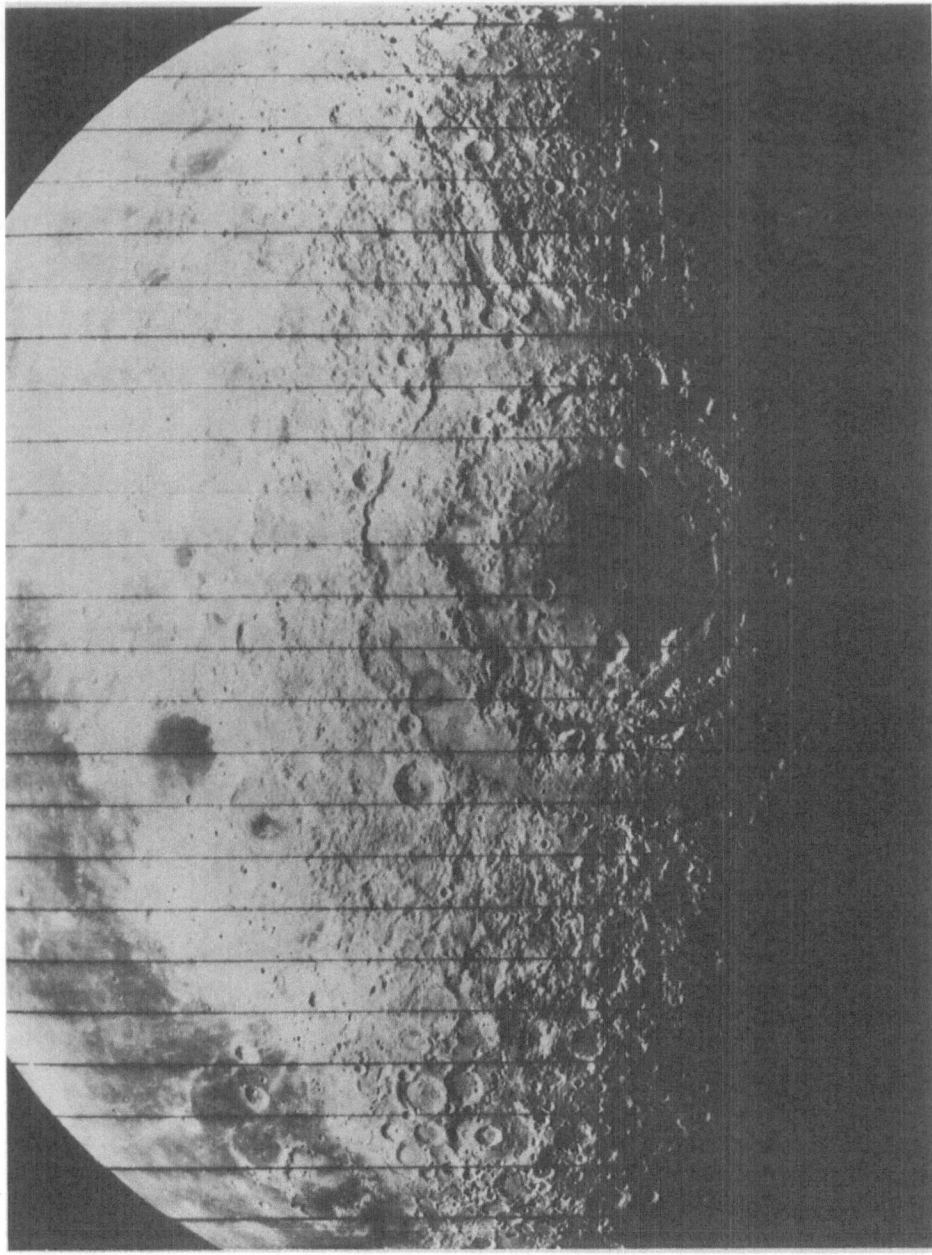

Fig. 1.29.   The lunar Mare Orientale – a huge triple-walled circular basin almost 1000 km across – as photographed by the Lunar Orbiter 4 on 25 May 1967 from an altitude of approximately 2700 km above the lunar surface (NASA-Langley photograph).

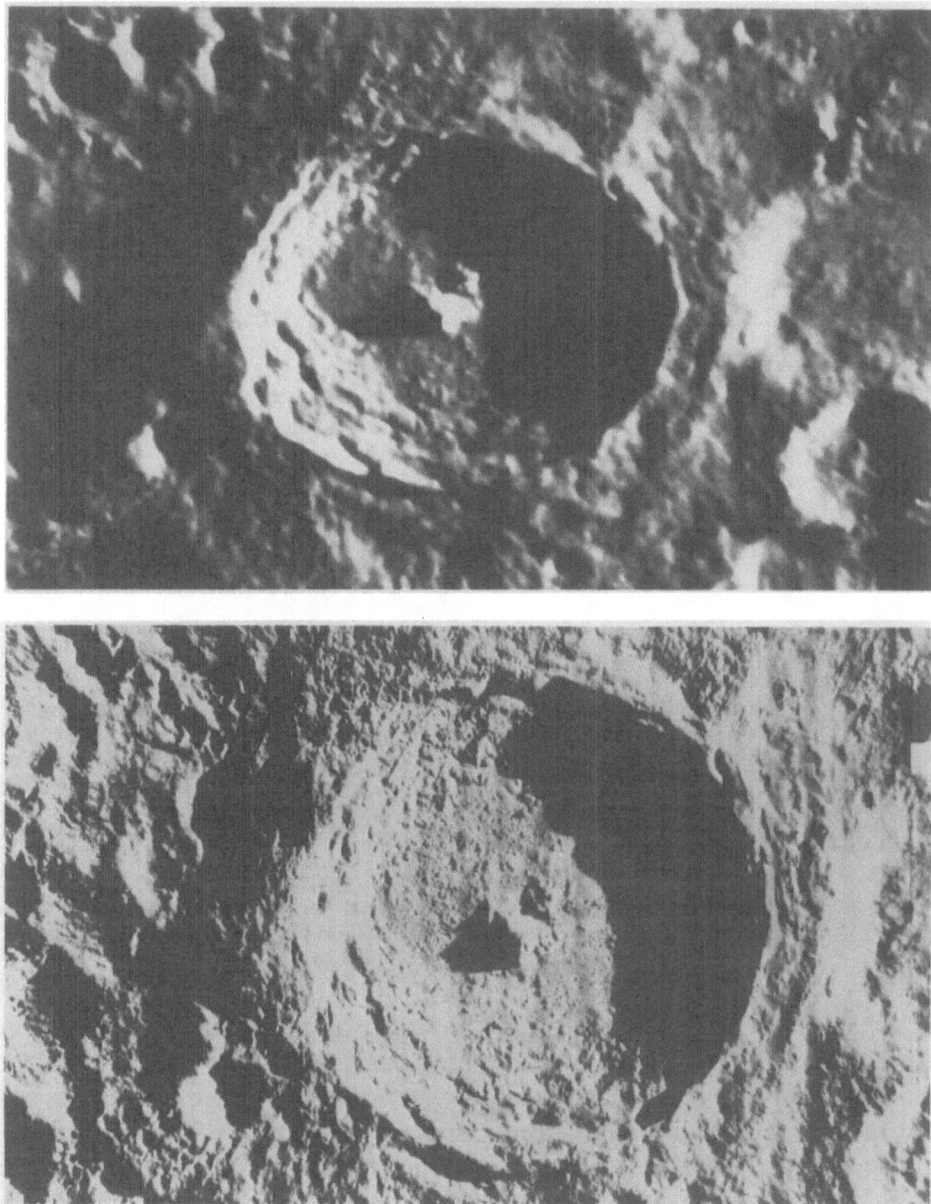

Fig. 1.30.   Another comparison of Earth-based and space-borne photography of the Moon – the crater
Tycho as photographed on 30 March 1966 with the 43-in. reflector of the Observatoire du Pic-du-Midi
(above), and by the Lunar Orbiter 5 on 15 August 1967 from an altitude of 105 km above the
lunar surface (below).

Fig. 1.31. The crater Tycho near the Moon's south pole, photographed by the Lunar Orbiter 5 on 15 August 1967 from an altitude of 105 km above the lunar surface. The area where Surveyor 7 landed in January 1968 is marked with a white circle; and for a view of that part of the lunar landscape from ground cf. Figure 1.18 (NASA-Langley photograph).

to a certain part of the Earth (the south-west part of the United States); and thus became briefly visible as a star of 12th apparent magnitude. This visibility lasted for more than one hour; and enabled the astronomers of the Catalina Observatory of the Lunar and Planetary Laboratory, University of Arizona, to secure its photographic record (Figure 1.35). Thus, under favourable conditions, a spacecraft of the size of the Orbiters can not only make itself 'heard' by radio signals transmitted with the power of a few watts from the distance of the Moon, but also actually 'seen' in specular reflection of sunlight at a proper geometrical configuration for time-intervals of the order of one hour.

Another very interesting result transpired from an analysis of extensive range-Doppler tracking of the motions of the lunar orbiting satellites, which between 1966–1968 described a total of over 6000 revolutions. Such satellites, in free flight, can also serve as passive sensors of lunar gravitational field; and thus, indirectly, on the distribution of lunar gravitational field. Muller and Sjogren (1968), in analyzing the orbital perturbations of the five U.S. Lunar Orbiters and constructing the first 'gravimetric map' of the Moon, discovered the existence of a localized sub-surface mass condensation ('mascons') which became an important tool for exploration of the interior of the lunar globe (cf. Chapters 4 and 5).

In contrast with the completeness with which all results obtained by the American Orbiters have promptly been made public, relatively little has so far been disclosed of the instrumentation of Lunas 10–12 and 14, or of the results obtained with their aid. Their payloads are, however, known to have been considerably larger and heavier than the Orbiters. The 245 kg Luna 10 was probably a selenodetic satellite sensing the gravitational field of the Moon, the results of which have at least partly been published (cf. Akim, 1966); and carried out further magnetic measurements in the proximity of our satellite (cf. Dolginov *et al.*, 1967). On the other hand, the 1640 kg

Fig. 1.32.   A high-resolution photograph of a part of the floor of the crater Tycho, taken by the Lunar Orbiter 5 at the same time (and from the same altitude) as the photograph reproduced on the lower part of
Figure 1.30. The area covered by this frame is approximately 11 × 13 km in size (NASA-Langley
photograph).

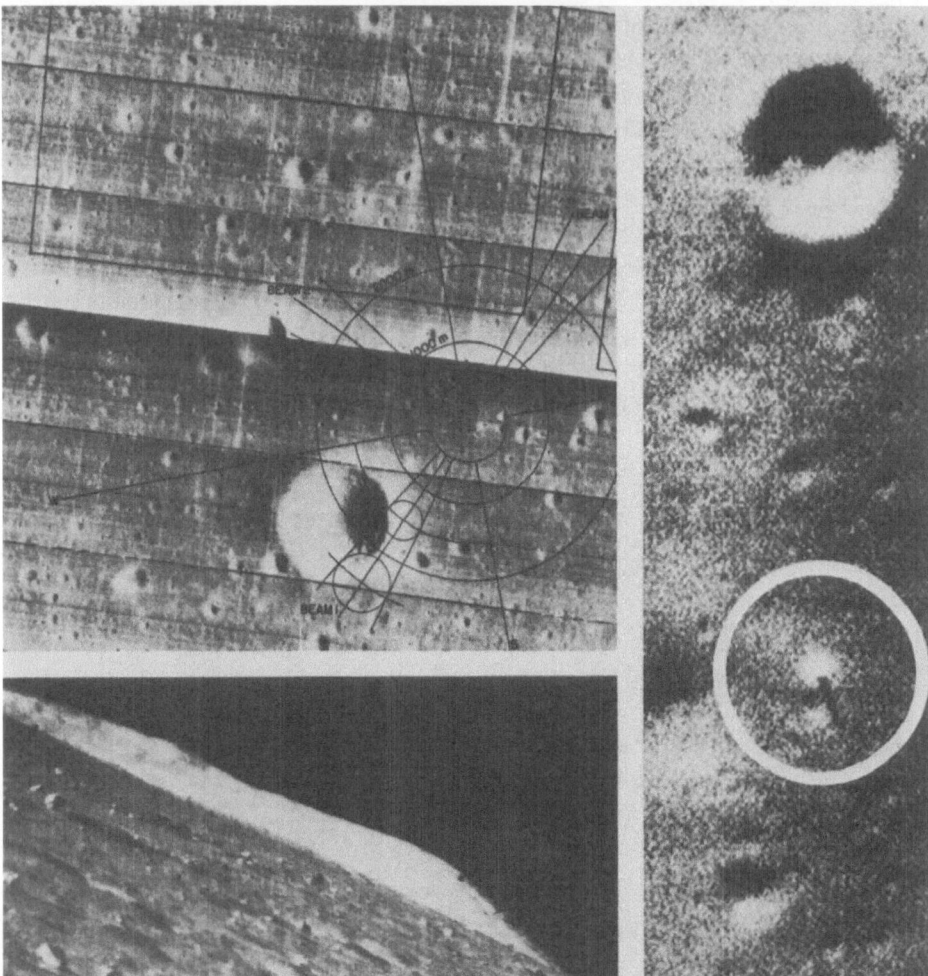

Fig. 1.33.   The landing place of Surveyor 1 in the Oceanus Procellarum, as photographed by the Lunar
Orbiter 1 in August 1966 (top left), and located by the Lunar Orbiter 3 in March 1967 from an altitude of
48 km (right). The photograph in the lower left corner shows ramparts of the 1-km crater televised by
Orbiter 1 from the vantage point marked on the photograph above it (NASA-Langley-JPL photographs).

Luna 11 was the heaviest spacecraft launched towards the Moon by 1966; and Lunas
12 and 14 were probably of a similar calibre. It has been announced that Luna 11
carried X-ray and γ-ray detectors among its instrumental complement; and also detectors of meteoroids; but very few results of their work have been published. Luna
12 was reported to have been on a photographic mission; but if so, no results of it
have again been made public; and nothing is known so far about the equipment or
fate of Luna 14. It is possible – though only conjectural – that the heavy Lunas 11, 12
and 14 may have been concerned primarily with the technology of re-entrant missions to the Moon. This, moveover, almost certainly true of the short-lived Lunas 15,
18 and 19.

There is, however, another satellite still revolving around the Moon on a scientific mission: namely, Explorer 35, launched on 19 July 1967 in a wide orbit of 684 min period. Explorer 35 – a Moon-anchored IMP only 104 kg in weight (cf. Figure 1.36). It carried no optics or photographic equipment; but its experimental repertoire includes magnetometers, plasma probes, energetic particle and cosmic dust detectors. Bi-static radar measurements of the electromagnetic properties of the lunar surface have been made by monitoring the transmitted and reflected RF signals. With their aid no evidence was, however, found in the records for any shock wave produced by the motion of the Moon in the interplanetary medium (through which the Moon – like the Earth – is moving with supersonic speed) – a fact bearing directly on the mean electrical conductivity of the lunar globe (see again Chapter 5).

And this brings us to a close of this brief survey of the new tools of lunar research

Fig. 1.34. The Ranger 8 impact area, photographed by the high-resolution lens of the Lunar Orbiter 5 in quest of the crater produced on the lunar surface by the impact of that had-landing spacecraft on 20 February 1965. It is believed that this crater is identical with one of the two marked by arrows on the print (cf. R. B. Baldwin, *Science* **157**, 546, 1967).

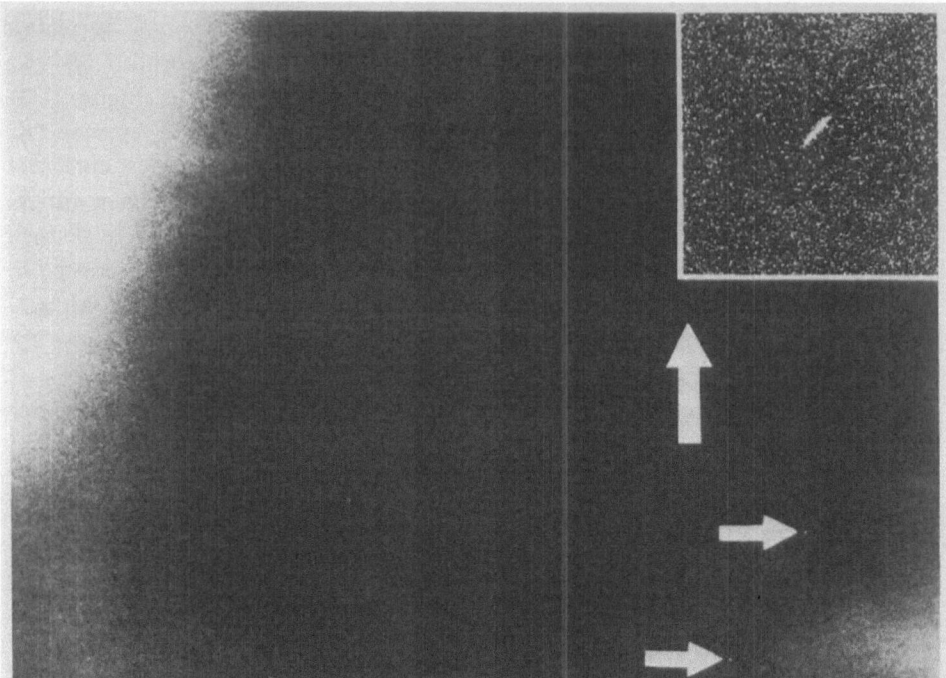

Fig. 1.35.　Lunar Orbiter 5 in orbit around the Moon as photographed from the Earth. Prior to the destruction of this spacecraft in January 1968, on command from the Earth, its solar batteries were oriented so as reflect sunlight to the Earth. When this occurred, the spacecraft became temporarily bright enough to be recorded on photographs taken with the 60-in. reflector of the Catalina Observatory of the Lunar and Planetary Laboratory, University of Arizona. On the left margin of this photograph we see a part of the overexposed limb of the Moon; while the inset on the right shows the spacecraft's motion (in enlargement) during the exposure (Reproduced by courtesy of the LPL, Tucson, Arizona).

which the students of the Moon acquired in the decade between 1959–1968, and the output of which thoroughly revolutionized our subject even before the first man set foot on the surface of our satellite on 20 July 1969.

On the human side, preceding this achievement which will be the subject of the next chapter, a glance at the data compiled in Table 1-1 reveals that all lunar research by means of spacecraft has so far been the exclusive domain of the Russians and the Americans, to which no other nation has so far been able to add any direct contribution (or is likely to make any for several years to come). It can, furthermore, be observed that, in all individual feats which have been accomplished up to 1966, the Russians achieved a priority: theirs was the first lunar fly-by (Luna 1 in January 1959), the first hard-impact (Luna 2 in September 1959) or circumnavigation (Luna 3 of October of the same year); the first soft-landing (Luna 9 in February 1966), as well as the first lunar orbiter (Luna 10 in March of the same year). However, the time-intervals of the Russian leads over the parallel American achievements have been progressively diminishing; and while it took more than two years after the Russian firsts for America to fly-by the Moon (Ranger 3 in January 1962) or to score a direct hit (Ranger 4 in April of the same year), the Russian lead-times for soft-landings or

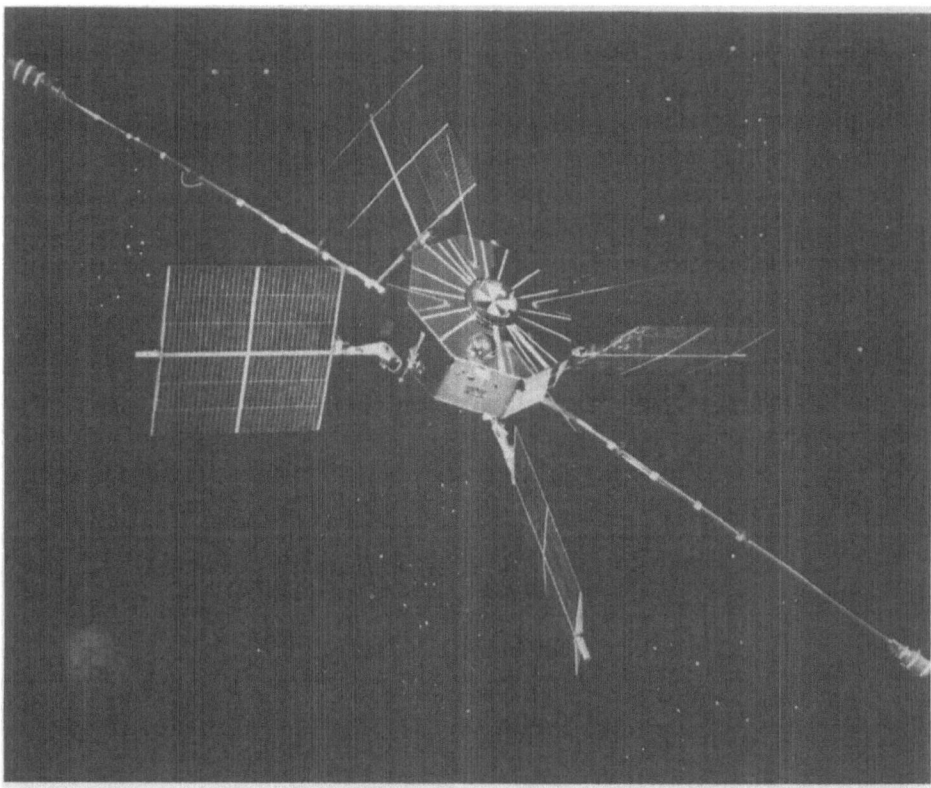

Fig. 1.36. The U.S. Explorer 35 spacecraft, with its solar panels and antennae outstretched (NASA official photograph).

injection of spacecraft into circumlunar orbits have been reduced to a few months. And what is more important – the American contributions, when they came, were on so massive a scale as to provide most part of the evidence we now possess. This is, in particular, true of direct photography of the lunar surface and of the mapping of the Moon based upon it (cf. Kopal and Carder, 1974); for in comparison with the tens of thousands of close-up photographs of the lunar surface provided by the U.S. Rangers, Surveyors and Orbiters, similar contributions of the Russian Lunas 9 and 13 or 18–19 – valuable as they may be in other respects – appear unimpressive. This is why so large a proportion of the pictorial evidence presented in this book has come from the American sources.

This disparity should, to be sure, in no way diminish the historical significance of the pioneer nature of much of the Russian space work; for its achievements paved the way; and once its feasibility became a proven fact, its follow-on was no doubt greatly encouraged by the knowledge that it can be done. In this sense, the Russian since 1957 pioneered the deep-space work just as the Americans pioneered the macroscopic use of nuclear energy after 1945. Without the Russian sputniks of 1957 there would undoubtedly not have been any National Aeronautics and Space Administration in the United States as we have it today; and without the Russian Moon probes of 1959, President Kennedy would have scarcely proclaimed in 1961 the Moon a target for the American astronauts in the latter part of the last decade.

The future historian of this subject will no doubt be impressed as we are by the fact that in order to be able to reach for the Moon in 1959, the Russian work on deep-space probes must have been started more than ten years before that time – in the difficult years when the ravages of the Second World War were still far from overcome. At that time, the only person who could have authorized the use to this end of the requisite means – both of material and manpower – was Joseph Stalin; and his courageous initiative contrasted favourably with a relative complacency with which the dawn of the space age in 1957 was greeted by President Eisenhower of the United States. But, perhaps, we should not praise – or blame – the politicians; but rather their respective advisors. Fortunately, the American nation as a whole reacted more vigorously to the challenge for science presented by the new endless frontier; with the results to which we shall turn more fully in the next chapter.

References

The prime reference to the work of the Russian Luna 3 is Barabashev, N. P., Mikhailov, A. A., and Lipski, Yu. N.: 1960, *Atlas of the Far Side of the Moon*, U.S.S.R. Acad. Sci. Moscow, English translation: Pergamon Press, London 1961.

The most important references on the work of the hard-landing U.S. Rangers are the following Technical Reports of the Jet Propulsion Laboratory, California Institute of Technology:

Rangers 3, 4, 5: JPL TR 32-199, 1962;
Ranger 6: JPL TR 32-605, 1964;
Ranger 7: JPL TR 32-700, 1965;
Ranger 8: JPL TR 32-800, 1966;
Ranger 9: JPL TR 32-800, 1966.

The prime reference to the work of the Russian Zond 3 is Lipski, Yu. N.: 1967, *Atlas of the Far Side of the Moon, II*, Izd. Nauka, Moscow. Concerning the accomplishments of the soft-landing Luna 9, cf. *The First Panoramas of the Lunar Surface etc.* edited by U.S.S.R. Acad. Sci., Izd. Nauka, Moscow, 1966.

The most important references on the work of the soft-landing US Surveyors are the following Technical Reports of the Jet Propulsion Laboratory, California Institute of Technology:

Surveyor 1: JPL TR 32-1023, 1966;
Surveyor 3: JPL TR 32-1177, 1967;
Surveyor 5: JPL TR 32-1246, 1967;
Surveyor 6: JPL TR 32-1262, 1968;
Surveyor 7: JPL TR 32-1264, 1969;
cf. also the Surveyor Project Final Report JPL TR 32-1265, 1969.

Preliminary reports on the individual Surveyors can also be found in the following Special Publications of the U.S. National Aeronautics and Space Administration:

Surveyor 1: NASA SP-126, 1966;
Surveyor 3: NASA SP-146, 1967;
Surveyor 5: NASA SP-163, 1967;
Surveyor 6: NASA SP-166, 1968;
Surveyor 7: NASA SP-173, 1968.

As regard the lunar orbiting satellites, the most useful references to their performance and accomplishments are the Contractor Reports to NASA by the Boeing Company, which appeared under the following numbers:

Orbiter 1: NASA CR-782 and 847, 1967.

Orbiter 2: NASA CR-883, 931, 1967; and CR-1141, 1968.
Orbiter 3: NASA CR-984, 1069 and 1109, 1968.
Orbiter 4: NASA CR-1054, 1092 and 1093, 1968.
Orbiter 5: NASA CR-1094 and 1142, 1968.

Of summarizing articles concerned with the accomplishments of lunar spacecraft, cf., e.g.,

Kopal, Z.: 1967, *Contemporary Physics* **8**, 331–356.
Jaffe, L. D.: 1969, *Space Sci. Rev.* **9**, 491–609.
Kopal, Z.: 1969, *Scientia* **104**, 3–32.
Jaffe, L. D., Choate, R., and Coryell, R. B.: 1972, *Moon* **5**, 348–367.

The following individual references have been quoted in this chapter:

Akim, E. L.: 1966, *Doklady U.S.S.R. Acad. Sci.* **170**, 799.
Arnold, J. R., Peterson, L. E., Metzger, A. E., and Trombka, J. I.: 1972, *Apollo 15 Prelim. Sci. Rept.* 16-1, NASA SP-289.
Dolginov, Sh. S., Yeroshenko, E. G., Zhuzgov, L. I., Pushkov, N. V., and Tyurmina, L. O.: 1960, *Iskustv. Sputniki Zemli* **5**, 149.
Dolginov, Sh. S., Yeroshenko, E. G., Zhuzgov, L. N., and Zhulin, I. A.: 1967, *Geomag. Aeronom.* **7**, 436.
Kopal, Z. and Carder, R. W.: 1974, *Mapping of the Moon*, D. Reidel Publ. Co., Dordrecht.
Lebedinsky, A. I., Aleshin, G. M., Iorenas, V. A., Krasnopolsky, V. A., Selivanov, A. S., and Zasetsky, V. V.: 1967b, in *Moon and Planets*, North-Holland Publ. Co., Amsterdam, pp. 65–70.
Muller, P. M. and Sjogren, W. L.: 1968, *Science* **161**, 680.
Sonett, C. P., Colburn, D. S., and Currie, R. G.: 1967, *J. Geophys. Res.* **72**, 5503.
Wattson, R. B. and Danielson, R. E.: 1965, *Astrophys. J.* **142**, 16.

CHAPTER 2

MANNED EXPLORATION: APOLLO (1969–1972)

The idea of visiting the Moon – in dream or reality – has flickered in the human mind since time immemorial. The entrancing visions of Kepler's *Somnium* (1634), or the professionally less competent excursions of Swift, Cyrano de Bergerac and of many others, resurrected for us more recently by Marjorie Nicolson (1948), belong properly to the domain of belles lettres. Listen, however, to the following extract of a letter which William Herschel – then at the beginning of his great career which led him to 'pierce the barriers of the heavens' – wrote on 12 June 1780 to the fourth Astronomer Royal, the Reverend Dr Nevil Maskelyne (cf. *Scientific Papers of Sir William Herschel*, Vol. I, pp. 40–41, London, 1912).

I beg leave to observe, Sir, that my saying there is almost an absolute certainty of the Moon's being inhabited, may perhaps be ascribed to a certain Enthusiasm which an observer, but young in the Science of Astronomy, can hardly divest himself of when he sees such wonders before him; and if you will promise not to call me a Lunatic, I will transcribe a passage ... which will shew my real sentiments on the subject.

Perhaps conclusions from the analogy of things may be exceedingly different from truth; but ... seeing that our Earth is inhabited and comparing the Moon with this planet: finding that in such a satellite there is a provision of light and heat: also, in all appearance, a soil proper for habitation fully as good as ours, if not perhaps better – who can say that is not extremely probable, nay beyond doubt, that there must be inhabitants on the Moon of some kind or another.

What a glorious View of the heavens from the Moon! How beautifully diversified with hills and valleys! No large oceans to take up immense plains fit for pasture, etc. Uninterrupted day on one half, and on the other a day and night of noble length, equal to many of ours! Do not all the elements seem at war here when we compare the Earth with the Moon? Air, Water, Fire, Clouds, Tempests, Vulcanos, etc: all these are either not on the Moon, or at least kept in much greater subjection than here....

For my part, were I to chuse between the Earth and Moon, I should not hesitate a moment to fix upon the Moon for my habitation.

We do not know what the kindly Reverend Maskelyne may have replied to such a letter; though some Astronomers Royal are known to have been very outspoken on this subject down to our own days. Recent advances in space technology, briefly surveyed in the preceding chapter, did expose the lack of vision of the more conservative members of the contemporary astronomical fraternity to an embarrassing extent; though Herschel's dreams on the attractions of life on the Moon ended up similarly as his views on the possibility of life on the Sun. Unmanned space work of the past decade prepared us to anticipate that no life – in any form – could await us on the Moon; though the 'glorious view of the heavens' beckoning from its surface continues to exert an attraction even more irresistible for the modern astronomer than even Herschel could have ever dreamt of.

The principal methods of unmanned exploration of the Moon through 1968 have already been described in the first chapter; and the aim of the present will be to introduce man to the stage at which he already played such a dramatic – albeit brief – role between 1969–1972 in the service of lunar exploration. In the United States of America, his role was foreshadowed by the historical pronouncement of President John F. Kennedy who on 26 May 1961, told Congress that "… this nation should commit itself to achieving the goal, before this decade is out, of landing a man on the Moon and returning him safely to the Earth" – a task which the combined ingenuity of American scientists and engineers succeeded in accomplishing by 1969 under the code name of Project Apollo. By its design as well as execution, this great project is bound to go down in history as one of the chief titles to glory of the past turbulent decade – and is sure to be remembered by posterity long after all more ephemeral events of recent past have been charitably confined to oblivion.

In the course of the seven Apollo manned missions undertaken between 1969–1972, no life was lost (or injury suffered) on trips whose combined mileage exceeded 7 million kilometres – a record rendering the Earth-Moon space communications by far the safest known way of man-travel. If, in spite of this enviable record, and of the epoch-making scientific accomplishments of these missions, the Apollo project came in 1972 to its premature halt, the reasons were not technical or scientific; but due merely to the fact that politicians less far-seeing than John F. Kennedy have since developed other and more pressing needs for the taxpayer's money.

To undertake a manned return trip to the Moon constitutes a task far more difficult than to operate in its proximity (or land on the Moon's surface) an unmanned spacecraft. By making the man a part of the system the requirement arises, in particular, for the provision of his life support which adds to the weight of the scientific apparatus and of the requisite fuel. In order to transport three men to the Moon on each mission, the Apollo planners had to await the completion of an advanced model of the Saturn C-5 booster built at NASA's Marshall Space Flight Center at Huntsville, Alabama, under the direction of Wernher von Braun. A schematic view of this giant three-stage rocket can be seen on the accompanying Figure 2.1; while Figures 2.2 and 2.3 shows the same configuration on the launching platform at Cape Kennedy, Florida, ready for a take-off to the Moon.

A few words of description of this Columbiad of the 20th century may be of interest to give in this place. Its first stage, 13 m in diameter, towers 51 m above ground; the second and third stages approximately 24 m in height (and $6\frac{1}{2}$ m in diameter) adding up to a total height of the assembled structure of close to 100 m. The total weight of this spacecraft – about 3000 tons at take-off – is equal to the combined weight of 25 Boeing 707 jet planes or (for those who prefer to think in naval terms) of one light cruiser.

To sail a cruiser at 30–40 knots through the seven seas of this Earth is one thing; but to lift it vertically and accelerate to a velocity in excess of 11 km s^{-1} (necessary to disengage it from the gravitational field of the Earth) is obviously another! In order to accomplish such a feat, vastly more powerful engines are clearly necessary. The first stage of Saturn C-5 rocket is lifted from the ground by a cluster of giant

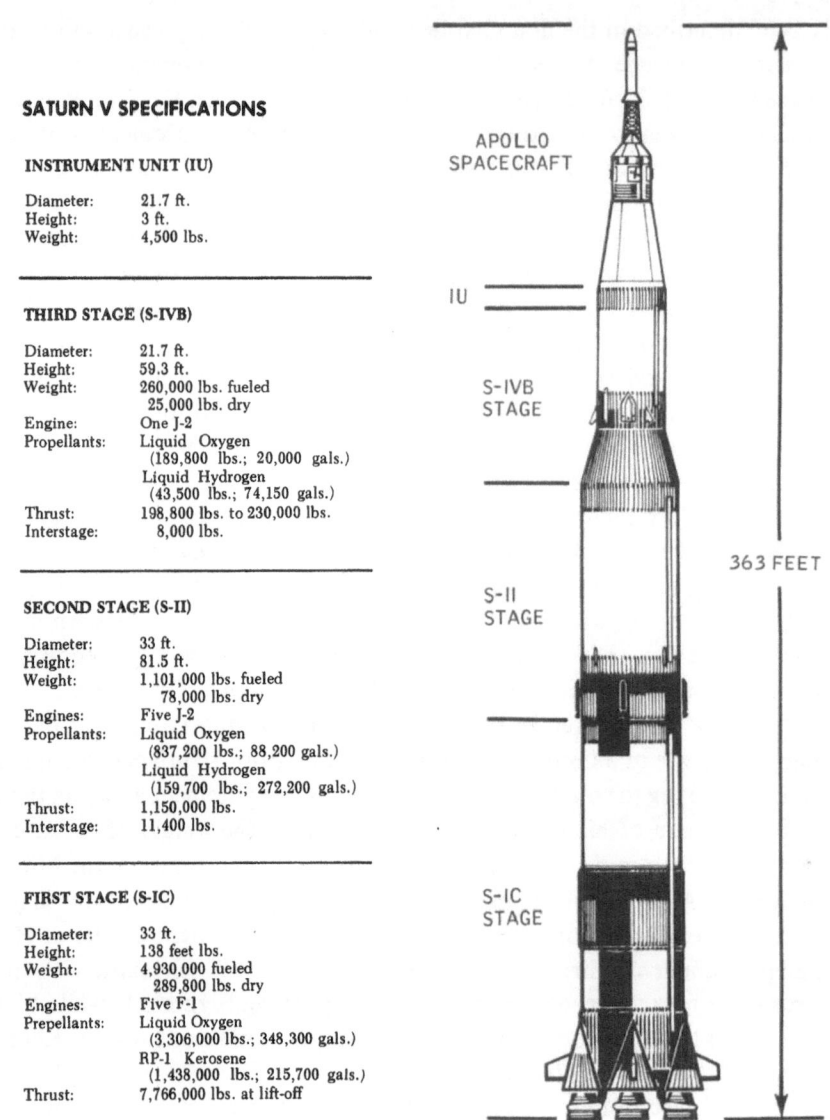

SATURN V SPECIFICATIONS

INSTRUMENT UNIT (IU)

Diameter:	21.7 ft.
Height:	3 ft.
Weight:	4,500 lbs.

THIRD STAGE (S-IVB)

Diameter:	21.7 ft.
Height:	59.3 ft.
Weight:	260,000 lbs. fueled
	25,000 lbs. dry
Engine:	One J-2
Propellants:	Liquid Oxygen
	(189,800 lbs.; 20,000 gals.)
	Liquid Hydrogen
	(43,500 lbs.; 74,150 gals.)
Thrust:	198,800 lbs. to 230,000 lbs.
Interstage:	8,000 lbs.

SECOND STAGE (S-II)

Diameter:	33 ft.
Height:	81.5 ft.
Weight:	1,101,000 lbs. fueled
	78,000 lbs. dry
Engines:	Five J-2
Propellants:	Liquid Oxygen
	(837,200 lbs.; 88,200 gals.)
	Liquid Hydrogen
	(159,700 lbs.; 272,200 gals.)
Thrust:	1,150,000 lbs.
Interstage:	11,400 lbs.

FIRST STAGE (S-IC)

Diameter:	33 ft.
Height:	138 feet lbs.
Weight:	4,930,000 fueled
	289,800 lbs. dry
Engines:	Five F-1
Prepellants:	Liquid Oxygen
	(3,306,000 lbs.; 348,300 gals.)
	RP-1 Kerosene
	(1,438,000 lbs.; 215,700 gals.)
Thrust:	7,766,000 lbs. at lift-off

APOLLO SPACECRAFT

IU

S-IVB STAGE

363 FEET

S-II STAGE

S-IC STAGE

Fig. 2.1.   A schematic view of the Saturn 5 rocket configuration, which between 1969–1972 carried seven manned missions to the Moon.

F1-jet engines, each of which provides a vertical thrust of 1 500 000 lbs. Their ensemble supplies, therefore, a combined thrust of 7 500 000 lbs. for a take-off – equivalent to almost 150 million standard horse power.

Needless to say, a delivery of this power requires the expenditure of a prodigious amount of chemical fuel. While its engines are burning (cf. Figure 2.3) the first stage of Saturn C-5 consumes close to 15 tons of LOX fuel (consisting of a mixture of two parts of kerosene to one part of liquid oxygen) per second of a flight lasting approximately $2\frac{1}{2}$ min – its total of 2250 tons of fuel used up during this time should be large enough to drive an average-size family car beyond the confines of the solar system! After the first stage has lifted the Saturn C-5 assembly off the ground through the main part of the atmospheric air mass to an altitude of approximately 100 km, the second stage equipped with five J-2 hydrogen-burning engines (of 1 million lb total thrust) takes over to place a 120-ton payload in an orbit around the Earth at an altitude of approximately 200 km. The third stage powered by a single J-2 engine will then accelerate some 45 tons of this payload to the Earth escape velocity and send it on its way to the Moon.

The journey to the Moon, during which (apart from a possible mid-course ma-noeuvre) the spacecraft whose schematic view is shown on Figure 2.4 remains in free flight, is likely to last 65–70 hr. After this time, when the lunar proximity has been reached, retro-rockets will decelerate the remaining spacecraft so as to enable it to be captured by the Moon in a closed cislunar orbit. In this orbit it may remain for many revolutions – until the time comes (depending on the choice of the landing site) when the remaining spacecraft separate: while the command and service module with one astronaut aboard continues to keep a lonely vigil in orbit around the Moon, the excursion module with two astronauts descends (with the aid of retro rockets of its own) to the selected region of the lunar surface (Figure 2.5). After the surface part

Fig. 2.2.   Saturn C-5 rocket at its launching platform on Cape Kennedy in Florida, being prepared for its mission to the Moon (NASA official photograph).

Fig. 2.3.   Anchors aweigh – take-off for a million-km trip to the Moon!

of the mission has been accomplished, the excursion module ascends again to regain an orbital altitude at which it can rendez-vous with the command module; and the three astronauts happily reunited then accelerate their remaining craft for the last time to its homeward journey back to the Earth.

This is not the place to give more detailed operational profiles of the successive nine Apollo missions which took off for the Moon between 1968–1972 and returned safely – beyond the salient facts listed already in Table 1-1, and augmented by those given in the accompanying Tables 2-1 and 2-2. The first two of such missions did not attempt yet to make any manned landings on the lunar surface – though their dates mark the time when the stage was set for wider exploits.

In particular, the date of December 24, 1968, will go down in history as a memorable landmark of human endeavour; for on that day three astronauts of the Apollo 8 mission, who 3 days before disengaged themselves from the gravitational field of the Earth, became temporarily attached to another celestial body. In other words, since

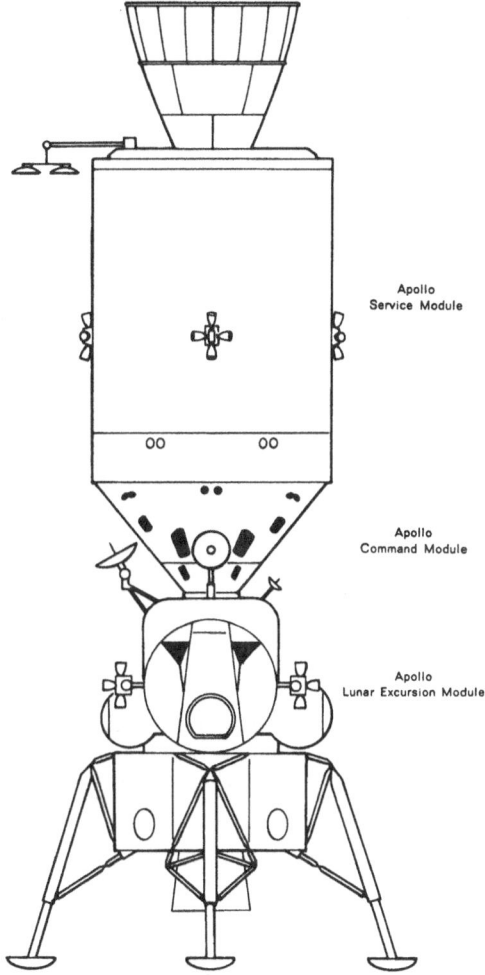

Apollo
Service Module

Apollo
Command Module

Apollo
Lunar Excursion Module

Fig. 2.4.   A schematic view of the Apollo spacecraft which carried men to the Moon nine times between 1969–1972. The upper part of the configuration (Service Module) served also as a fuel-storage tank. The middle part (Command Module) provided the living quarters for the three astronauts for most part of their long journey; and continued to orbit the Moon with one astronaut aboard when the other two used the Lunar Excursion Module (below) to descend to the surface, and to return back with a part of it after using its stand as a launching platform on the Moon.

Fig. 2.5. The Lunar Excursion Module (LEM) 'Eagle' of the Apollo 11 mission on the lunar surface in Mare Tranquillitatis on 20 July 1969. Astronaut Neil Armstrong – the first man to set human foot on another celestial body – can be seen in front of the LEM with life-support system in his back-pack (NASA official photograph).

Christmas 1968 *Homo Sapiens* – species born and bred on this planet, ceased to inhabit the Earth alone, and took the first tentative steps towards his proliferation throughout the solar system. For during the 20 hr when they circled around the Moon, these astronauts became temporary denizens of the Moon in the same sense as the Earth-circling astronauts are terrestrials. The time spent was extended to three days in volunatry lunar captivity by the astronauts of the Apollo 10 mission in May 1969. But it was not till the memorable July mission of Apollo 11 that spacecraft carrying two men actually descended on the lunar surface, to inaugurate the era of an even closer intimacy with our nearest celestial neighbour.

TABLE 2-1

Place and time of re-entrant spacecraft landing on the Moon

Spacecraft	Place of landing			Date and time of landing	
	Region	Longitude	Latitude		
Apollo 11	Mare Tranquillitatis	23°29′24″ E	0°40′12″ N	1969 July 20	6^{h}14^m
Apollo 12	Oceanus Procellarum	23°20′23″ W	2°27′ 0″ S	1969 November 18	20 52
Luna 16	Mare Foecunditatis	56°18′ E	0°41′ S	1970 September 20	5 18
Apollo 14	Fra Mauro	17°27′55″ W	3°40′24″ S	1971 February 5	8 37 10
Apollo 15	Mare Imbrium	3°39′10″ E	26° 6′ 4″ N	1971 July 30	22 16 29
Luna 20	Mare Foecunditatis	56°30′ E	3°34′ N	1972 February 21	19 19
Apollo 16	Descartes	15°30′47″ E	8°59′34″ S	1972 April 21	2 23 36
Apollo 17	Taurus-Littrow	30°45′26″ E	20°9′41″ N	1972 December 11	19 54 57

TABLE 2-2

Manned flights to the Moon (1968–1972)

Mission	Date of the start of the mission	Participating astronauts *	Duration of the entire mission	Duration of stay on the lunar surface
Apollo 8	1968 Dec 21	F. Borman, J. A. Lovell, W. A. Anders	147^{h}0^m	–
Apollo 10	1969 May 15	T. P. Stafford, J. W. Young, E. A. Cernan	192 3	–
Apollo 11	1969 July 16	N. A. Armstrong, M. Collins, E. E. Aldrin	195 18	21^{h}36^m
Apollo 12	1969 Nov 14	Ch. Conrad, R. F. Gordon, A. L. Bean	244 36	31 29
Apollo 13	1970 Apr 11	J. A. Lovell, J. L. Swigert, F. W. Haise	142 55	–
Apollo 14	1971 Jan 31	A. B. Shepard, S. A. Roosa, E. D. Mitchell	215 21	34 11
Apollo 15	1971 July 26	D. R. Scott, A. M. Worden, J. B. Irwin	295 12	66 55
Apollo 16	1972 Apr 16	J. W. Young, T. K. Mattingly, Ch. M. Duke	290 36	72 58
Apollo 17	1972 Dec 7	E. A. Cernan, R. E. Evans, H. H. Schmitt	304 41	75 1

* The first name is that of the Commander of the respective mission; and the second, that of the Command Module Pilot (i.e., orbiting astronaut).

The more human aspects of this historical encounter have been written up in numerous books in many languages which appeared in the literature since 1969. In what follows we shall, therefore, confine our main attention on the scientific equipment which the successive Apollo missions deployed on the Moon, and whose combined output largely furnished the data discussed in subsequent chapters of this book. In doing so, we shall de-emphasize the role of scientific apparatus which merely 'hitch-hiked' a ride to the Moon for a study of objectives not directly concerned with the Moon – such as the Sun and its radiative as well as corpuscular output (solar wind, cosmic rays), the interplanetary medium (micrometeorites), or the gravitational waves and other aspects of galactic studies – not because the results of such experiments are any less interesting; but because their connection with the Moon is largely incidental and, therefore, their closer discussion is outside the scope of this book.

As far as the lunar studies are concerned, the primacy in scientific contributions secured by instruments installed and operated on the lunar surface should probably go to the seismic experiments – both passive and active – which constituted an essential part of the Apollo 11–16 missions, and the outcome of which contributed very largely to our present knowledge of the state of the interior of the lunar globe, as exposed in Chapter 5 of this book.

The seismometers installed on the surface of the Moon (cf. Figure 2.6) consist – like their terrestrial prototypes – basically of mass free to move in our direction, suspended by means of a spring (or a combination of springs and hinges) from a framework that is firmly connected with the lunar surface and shares its motion (if

Fig. 2.6.   Apollo 11 astronaut Edwin Aldrin deploys the passive seismic experiment package on the surface of the Moon. The cylinder in the middle of the package houses the seismometer, flanked on both sides by two arrays of solar cells to provide the electrical energy for operation and transmission (NASA official photograph).

any), while the suspended mass tends to remain fixed in space because of its own inertia. The resulting relative motion between this mass and its framework can be recorded and telemetered from the Moon to the Earth, where it can be used to calculate the lunar ground motion from known properties of the instruments.

The seismometer installed on the Moon by Apollo 11 ceased to function two months later due to instrumental reasons; but all others are still operative; and with the successful installation of one at the Descartes landing site of Apollo 16 in April 1972, a 4-station lunar seismic network was completed, spanning the near face of the Moon in the pattern of nearly-equilateral triangles with an average spacing of 100 km between individual stations. Since the arrival on the Moon of Apollo 15 in July 1971, the combined output of this network makes it possible to identify the location of each seismic disturbance within the lunar globe uniquely. The results obtained from the data supplied by this network so far will be reviewed in Chapter 5; and there is no room for doubt that a continuing operation of the Apollo seismic network for many years to come will furnish many more data bearing on the gross structure of the lunar globe as well as of those caused by meteoroid impacts on the lunar the Earth-Moon system impinging on the lunar surface.

In addition to the passive records of seismic events, due to the internal activities of the lunar globe as well as of those caused by meteoroid impacts on the lunar surface, the seismographs on the Moon recorded also the disturbances produced by active (i.e., man-made) experiments – involving planned detonation of explosives in the proximity of the landing site (Apollo 14 and 16), or impacts on the Moon of the excursion modules of the spacecraft which could be discarded after the final rendez-vous of the three astronauts in a cislunar orbit, and directed to crash-land on the Moon.

A study of the active (man-made) seismic disturbances can be made to comple-ment the results deduced from the passive (natural) in two ways: namely, in the scale of the phenomena under investigation, and the total amount of energy released by them. Passive seismic events, occurring generally at great depths, lend themselves for the study of the internal structure of the lunar globe as a whole; while man-made surface events have been designed to study primarily the structure of the layers in the proximity of the respective landing site. In addition, the energy release in man-made experiments is known in advance, and does not have to be treated as an ad-ditional unknown. The impacts of the S-IV B stages of used-up spacecraft provided (on account of their relatively large mass and high velocity) the most energetic seismic events observed on the Moon so far.

Another scientific instrument which went up to the Moon already with Apollo 11 in July 1969, and again with Apollos 14 and 15 in February and July 1971, was a cube-corner reflector for the return of laser pulses beamed on it from the Earth. As is well known, a light beam incident on the corner of a cube is reflected successively from its three faces and returned in a direction exactly parallel with that of the in-cident beam – whatever its direction may have been. The only kind of light sent out from the Earth that has any chance whatever to return a measurable 'echo' is represented by short pulses of radiation possessing coherence in phase (laser); and the parallelism between the incident and reflected beam ensures that the laser pulse reflected from the Moon will return to the vicinity of its origin on the Earth.

The Apollo 11 and 14 laser retro-reflectors (see Figure 2.7) consisted of 100 cube-corners made of fused silica, with front face diameters of 3.8 cm; while the Apollo 15 retro-reflector consisted of 300 such cube-corners. Their role on the Moon has been to return to the Earth measurable echoes of the laser pulses sent out to them by means of large astronomical reflectors from different observatories of the world (Lick, McDonald, Pic-du-Midi); and accurate timing of a time-lag between the outgoing pulse and its returning echoes added two or three orders of magnitude to the exactitude with the Earth-Moon distance (or, more precisely, that between the trans-

Fig. 2.7.   The lunar retro-reflector of the Apollo 14 mission, with its array of 100 cube-corners for returning back to the Earth the terrestrial laser pulses (NASA official photograph).

mitting (receiving) facility on the Earth and the retro-reflector on the Moon) can be established from the measurements (for fuller details see the following Chapter 3). To give parallel unmanned lunar research its dues, however, let us recall that cube-corner retro-reflectors of French design, and of somewhat larger aperture, were transported to the Moon also by the unmanned Russian Lunokhods (Luna 17 and 21).

In addition to the instruments required for seismic work and laser ranging, the Apollo Lunar Surface Experiments Package (ALSEP; see Figure 2.8) installed on the Moon by the missions 12 and 14–17 included magnetometers to measure the local magnetic fields on the surface of our satellite; cold cathode ion gage and supra-thermal ion detector for studies of the particulate contents of lunar environment (see Chapter 9); in addition to various devices to study the heat flow through sub-surface

Fig. 2.8. The Apollo lunar science experimental package (ALSEP) of Apollo 14 mission *in situ* on the Moon (NASA official photograph).

Fig. 2.9.   Lunar field geology – astronaut-scientist H. H. Schmitt of the Apollo 17 mission at work in
December 1972 in the neighbourhood of the crater Littrow (NASA official photograph).

layers (Apollo 15 and 17); the electrical properties of the lunar surface (Apollo 17);
and the mechanics of lunar soils (Apollo 14–17).

Needless to say, ample attention was devoted on each mission to lunar field geology
(Figure 2.9) and photography – an effort in which the astronauts of the last three
missions (Apollo 15–17) were greatly assisted by the availability of lunar roving
vehicles (Figure 2.10) which enabled them to extend their range of exploration far
beyond the radius of action of the first three manned missions (Apollo 11, 12 and 14).

Our account of the research programmes carried out on the Moon by the six
manned Apollo missions has so far been limited to activities on the lunar surface
itself. But only two astronauts of each mission actually descended to the ground;
while the third remained in the command module (cf. Figure 2.11) in a circum-lunar

Fig. 2.10. Man and his vehicle – the lunar rover manned by Astronaut D. R. Scott of the Apollo 15 mission at the foot of the lunar Apennines (Mt. Hadley, the highest peak of this chain of mountains, in the background) (Official NASA photograph).

orbit in anticipation of the final rendez-vous. On the first two missions (Apollo 11 and 12) – and partly also on the third (Apollo 14) the astronaut in orbit was the 'forgotten man' of the project, having little to do and knowing less of what went on on the lunar surface in the meantime than the countless spectators at their television sets on the Earth. It was not till with the Apollo 15 mission – the one which saw the importation of lunar rover to the surface – that the situation changed radically also in orbit, and the orbiting astronaut became at last a full-fledged partner in lunar exploration as well.

The 'orbital science' made its first contributions to lunar studies with Apollo 12 – in the form of multicolour photography of the Moon, followed by a photography

Fig. 2.11. The Lunar Orbiting Module of the Apollo 17 mission with its SIM-bay exposed (NASA official photograph).

of faint diffuse celestial objects (gegenschein) from Apollo 14. It was, however, not till for the last three missions (Apollo 15–17) that carried two different systems of specially designed optics operated automatically to obtain high-resolution panoramic photographs of the Moon's surface overflown by the orbiting spacecraft, and high-quality metric photographs to study the topography of this surface along the ground track.

The panoramic camera, equipped with a lens of 60.96 cm focal length, rotates continuously in a direction across the path of the orbiting spacecraft (in order to provide for the panoramic scanning); and also tilts backwards and forwards to provide for a stereoscopic coverage. In addition, in order to prevent blurring of the images during exposures the camera automatically compensates for the forward motion of the spacecraft. Lastly, a special sensor detects the ratio of the forward velocity to the altitude of the camera above the lunar surface, and automatically corrects for it. As a result of all these corrections, the panoramic camera can record on its 5-in. wide strips of films details a metre or less in size from an orbiting altitude of 100 km – a no mean achievement of space photography.

The other ('metric') camera carried by the orbiting command module consists of two photographic objectives of 7.62 cm focal length pointing in exactly opposite directions: one photographs the lunar landscape underneath the spacecraft, while the other records at the same time the star field above which should permit the exact orientation of the camera to be determined later on. The linear resolution attainable with this camera is not more than about 20 m on the lunar surface; but the field distortion is extremely small, and individual points can be located on the films and related to each other, within the 20-m accuracy, over very large areas of the lunar surface.

Both the 'metric' and 'panoramic' cameras are operated automatically as a part of the scientific instruments bay (SIM-bay) of the command module (see Figure 2.11); and furnished several thousand photographs on each mission. A re-loading of the used films by fresh magazines constitutes a part of the extra-vehicular activity for the astronaut in orbit (Figure 2.12). Photographs which have been secured by these cameras are of unsurpassed quality, and represent the best documents on the to-pographic structure of the lunar surface which we possess so far – as the reader can judge for himself from the examples reproduced on Figures 2.13–2.16 of this chapter, and elsewhere in this book.

In addition to the equipment for high-resolution lunar photography, the SIM-bays of Apollos 14–17 carried several instruments designed to enable us to study the structure of the lunar surface in depth. One was the S-band Transponder – a radio device which receives from the Earth (while the spacecraft is in view) waves of a very steady frequency of 2115 Mc s^{-1}, and re-transmits back to the Earth, subject to a Doppler shift in frequency arising from the relative motion of the orbiting spacecraft. Since the frequency of the returning waves can be measured with a precision better than one part in 10^5, the S-band transponder enables us to keep accurate track of the radial velocity of the orbiting spacecraft, modulated by the lunar gravitational field. It is primarily with the aid of such a device that Muller and Sjogren discovered in 1968 the existence of geographically localized gravitational anomalies on the Moon, associated with sub-surface mass-condensations on which more will be said in Chapter 4.

Bi-static radar investigation (Apollo 14–16) constitutes another experiment which constitutes a by-product of the radio communications between the Earth and the orbiting command module. Signals coming directly from the spacecraft are compared with their reflections from the lunar ground below; and observed intensity differences can be utilized to study the dielectric properties and average roughness of the lunar ground.

The Sounder Experiment of Apollo 17 represented a further step in the explora-

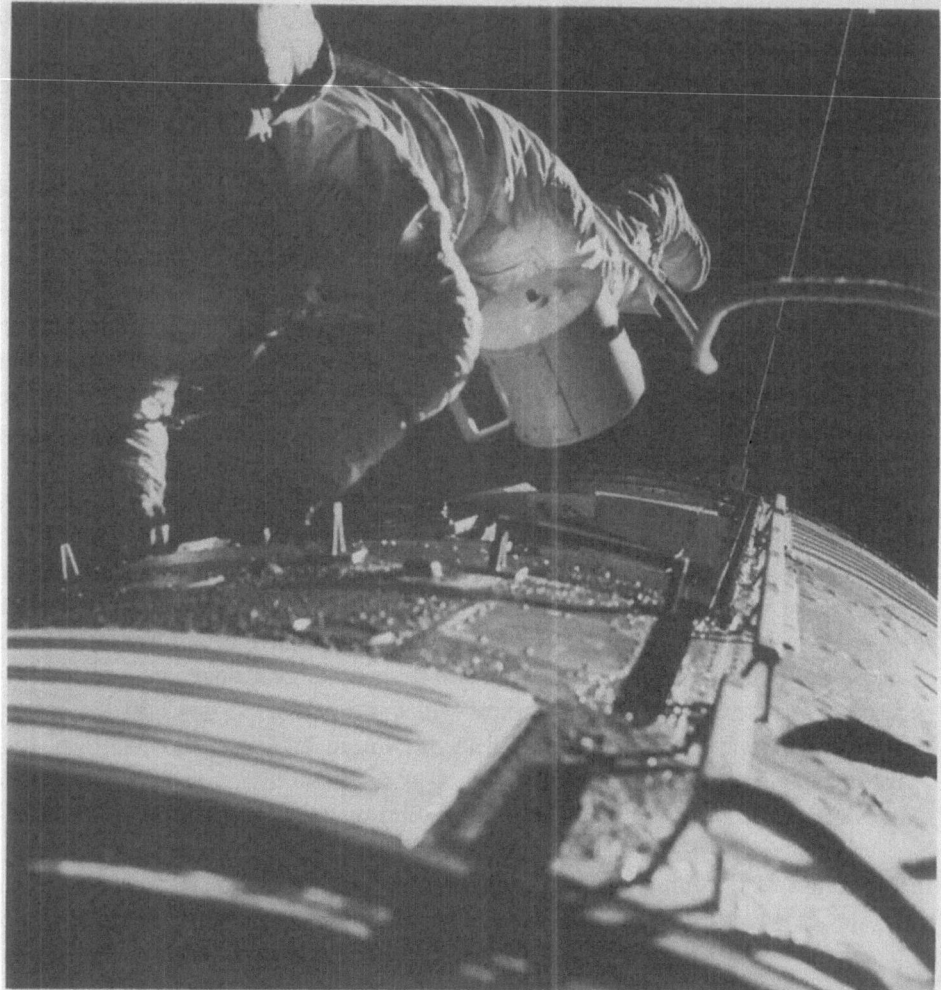

Fig. 2.12.   Astronaut R. Evans, Command Module Pilot of the Apollo 17 mission, re-loads the film magazines of his automatic cameras in the SIM-bay (NASA official photograph).

tion of the lunar surface in depth by means of electromagnetic waves. Radar signals at three different frequencies (5, 15 and 150 Mc s^{-1}) are sent out by the command module and back-scattered by the lunar surface. On account of the relatively low frequencies of such signals, they can penetrate deep into the lunar ground and be back-scattered in layers deep below the visible surface. The character of the reflected signals has told us much about the degree of fragmentation of the layers constituting the lunar surface, and on the dielectric properties of their material, down to a considerable depth.

Another important scientific instrument contained in the SIM-bay of Apollo 15–17 missions has been a laser altimeter – a device to measure the vertical deformations of the lunar surface along the ground track overflown by the respective spacecraft.

Fig. 2.13. An example of lunar photography from the orbit – this photograph of the craters Goclenius and Magelhaens was taken by the Apollo 8 mission in December 1968, has been a part of the first shipment of first-generation lunar negatives from the lunar vicinity to the Earth (NASA official photograph).

The principle of this instrument is quite similar to that employed in tracking the Moon's motion from the Earth, and described already in an earlier part of this chapter. In the present case, the emitting laser source is placed on the orbiting command module, and coherent light pulses sent out to illuminate the sub-spacecraft point.

On account of the much closer proximity of the surface thus illuminated, a diffuse reflection from it will give rise to a measurable light signal without the need of any artificial devices on the ground; and the time-lag between the outgoing signal and its returning echo can then be converted into the absolute distance between the transmitting source and the reflecting element of the lunar surface with an error of less than one metre. Since, moreover, the position of the transmitter in space is known to us with at least the same accuracy, the data furnished by the laser altimeter can

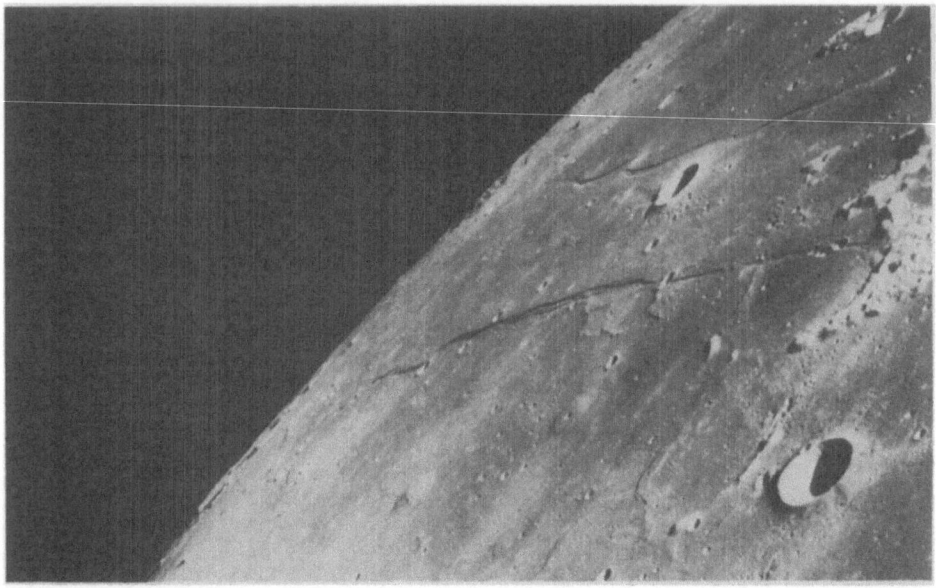

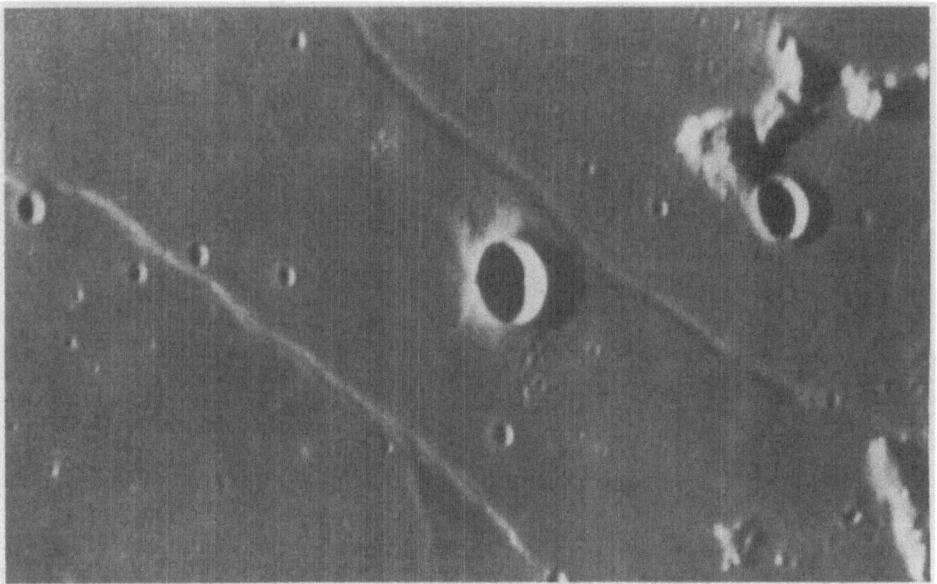

Fig. 2.14.   A photograph of the plains of Mare Tranquillitatis on the Moon, showing the crater Cauchy in between a shallow rille and a low left ('Cauchy's hyperbolae'), with a typical dome to the north of the cleft. Photograph taken by the Apollo 8 mission in orbit around the Moon on 24 December 1968 (above) should be compared with a terrestrial photograph (below) of the same region, taken with the 43-in. reflector of the Observatoire du Pic-du-Midi (Manchester Lunar Programme).

give us the detailed shape of the overflown ground track to a precision exceeding by a factor of 100 that previously attainable by stereogrammetric methods from the distance of the Earth. This has indeed been accomplished; with the results to be described more fully in Chapter 4.

In addition, the complement of scientific instruments aboard the orbiting spacecraft of Apollo 15 and 16 included an X-ray, α-particle and γ-ray spectrometers, as well as a mass spectrograph: the X-ray fluorescence spectrometer to measure the flux of X-rays produced on the lunar surface by impinging particles of the solar wind; α- and γ-ray spectrometers to measure the natural radioactivity of the lunar surface; and a mass spectrometer to measure the masses of the atomic constituents of the

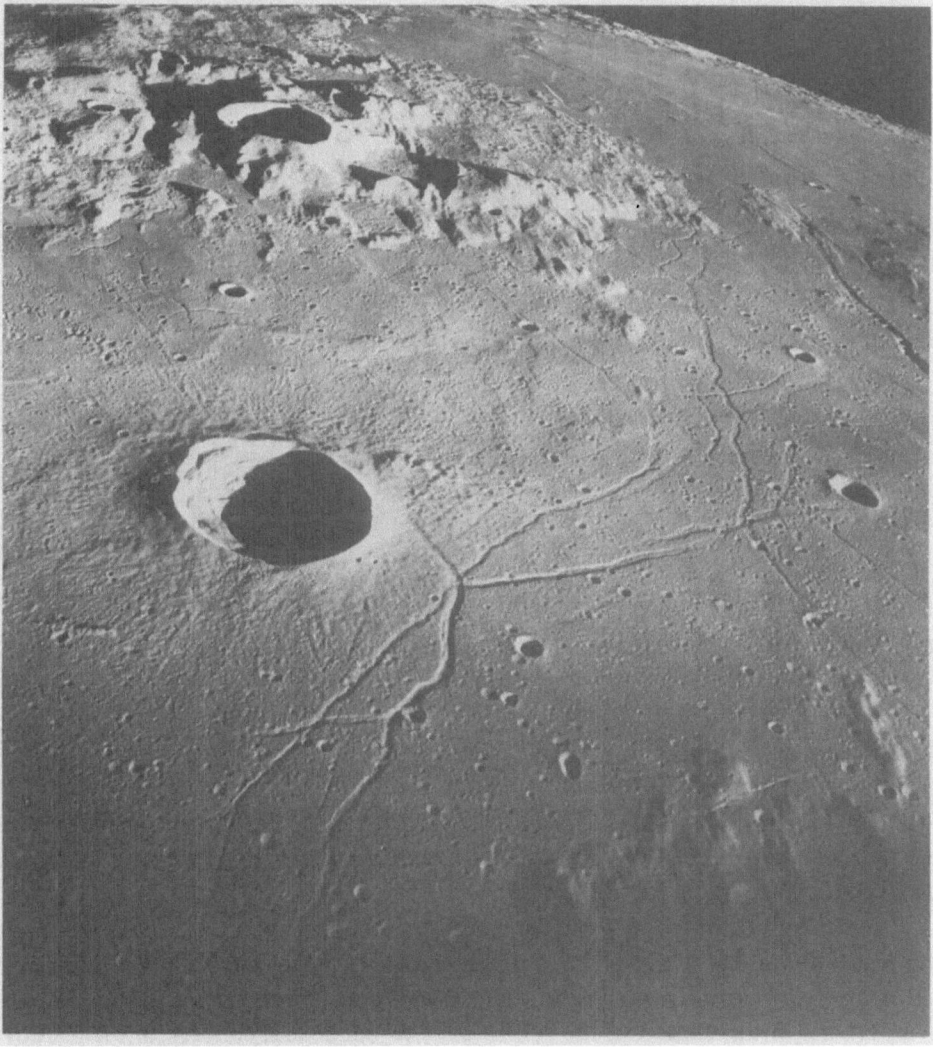

Fig. 2.15. A photograph of the crater Triesnecker and its rilles in the plains of Sinus Medii from the orbiting spacecraft of Apollo 10 mission in May 1969 (NASA official photograph).

Fig. 2.16.   A photograph of the crater Copernicus (cf. also Figure 1.27), taken from the Command Module of Apollo 12 mission in November 1969 (NASA official photograph).

lunar exosphere. The results deduced from the X-ray spectra on the chemical composition of the lunar crust, as well as the outcome of the natural radioactivity studies will be summarized in Chapter 7; while the constitution of the lunar exosphere based on the mass spectrometry of Apollo 15–16 as well as on the UV-spectroscopy aboard Apollo 17 will be discussed in Chapter 9.

It remains to be noted that the command modules of Apollo 15 and 16 missions were not orbiting the Moon alone, but were accompanied by sub-satellites (cf. Figure 2.17) carrying scientific instrumentation of their own – instrumentation concerned with studies of the motion of the sub-satellite (S-band transponder), of the shadow cast by the Moon in the solar wind, and with the measurements from the orbit of the lunar magnetic field.

And by this our brief account of the principal instrumental equipment of the Apollo missions between 1969–1972 has come to the end. Looking in retrospect at their accomplishments, we perceive that the entire Apollo effort can be divided in three groups: Apollo 7–10 missions which gradually paved the way for manned landings from the Earth to the Moon; Apollo 11–14 missions which initiated our acquaintance with the lunar surface; and Apollos 15–17 which carried out most of the actual scientific objectives of the whole programme. The fact that their follow-on was cut short of the three additional missions originally planned must be regarded as a veritable tragedy of historical proportions; with those responsible for it probably unaware of the severity of the verdict by which posterity is bound to condemn them for their lack of judgment.

It remains only to be added that while the American astronauts visited the surface of the Moon no less than six times between 1969–1972 and returned home in triumph, the Russians did not stay idle. Even before the Apollo 8 paid an orbital visit to the Moon on Christmas Eve 1968, the unmanned Russian spacecraft Zond 5 and 6 successfully accomplished similar astronautical manoeuvres in September and October of the same year, and returned with photographs such as that shown on the accom-

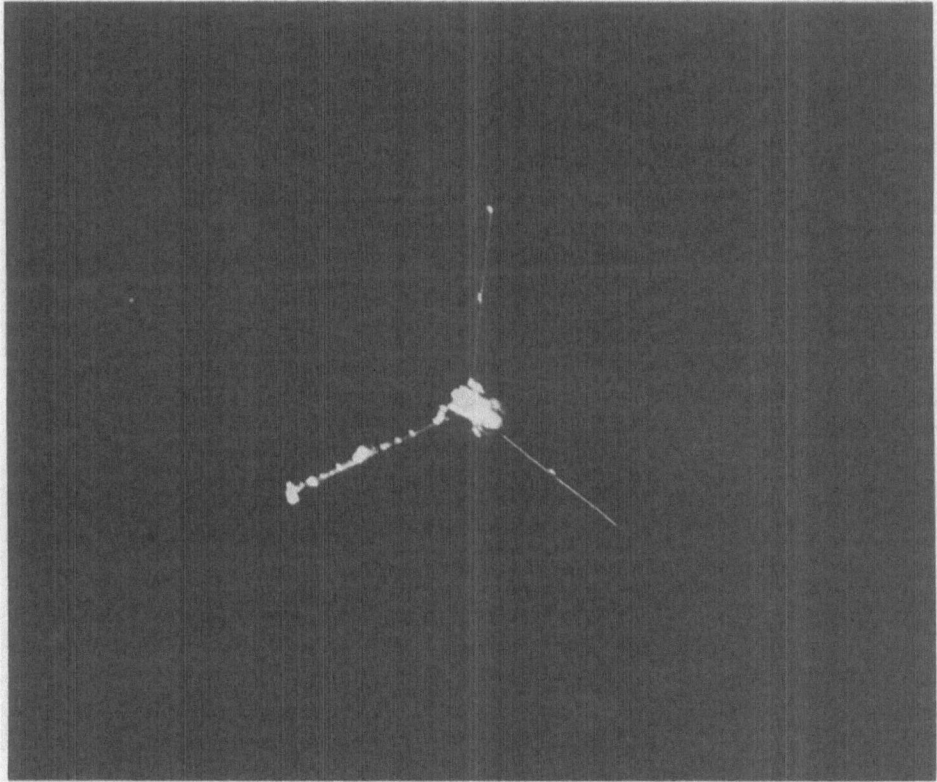

Fig. 2.17.   A sub-satellite of the Command-Service Module of Apollo 15 mission in its orbit around the Moon (NASA official photograph).

Fig. 2.18.   A photograph of the twin craters Vavilov on the far side of the Moon (cf. also Figure 1.4), taken and brought back to the Earth by the U.S.S.R. spacecraft Zond 6 in November 1968 (U.S.S.R. Acad. Sci. photograph).

panying Figure 2.18. But towards the end of the 1960's it had become clear that the U.S.S.R. will stay clear of the 'manned race' for the Moon, and has concentrated its material as well as human resources on the development and operation of recoverable unmanned spacecraft capable of soft landings on the Moon, followed by a return to the Earth with samples of lunar material scooped up at the landing place. The first prototype of this series of spacecraft which was successfully injected into a cislunar orbit was probably Luna 15 of July 1969 (cf. Table 1-1); and only a technical mishap may have prevented this spacecraft from returning to the Earth with the first sample of lunar rocks collected automatically before the return of Apollo 11 – a feat accomplished by Luna 16 in September 1970 and Luna 20 in February 1972. A schematic view of this spacecraft can be seen on the accompanying Figure 2.19.

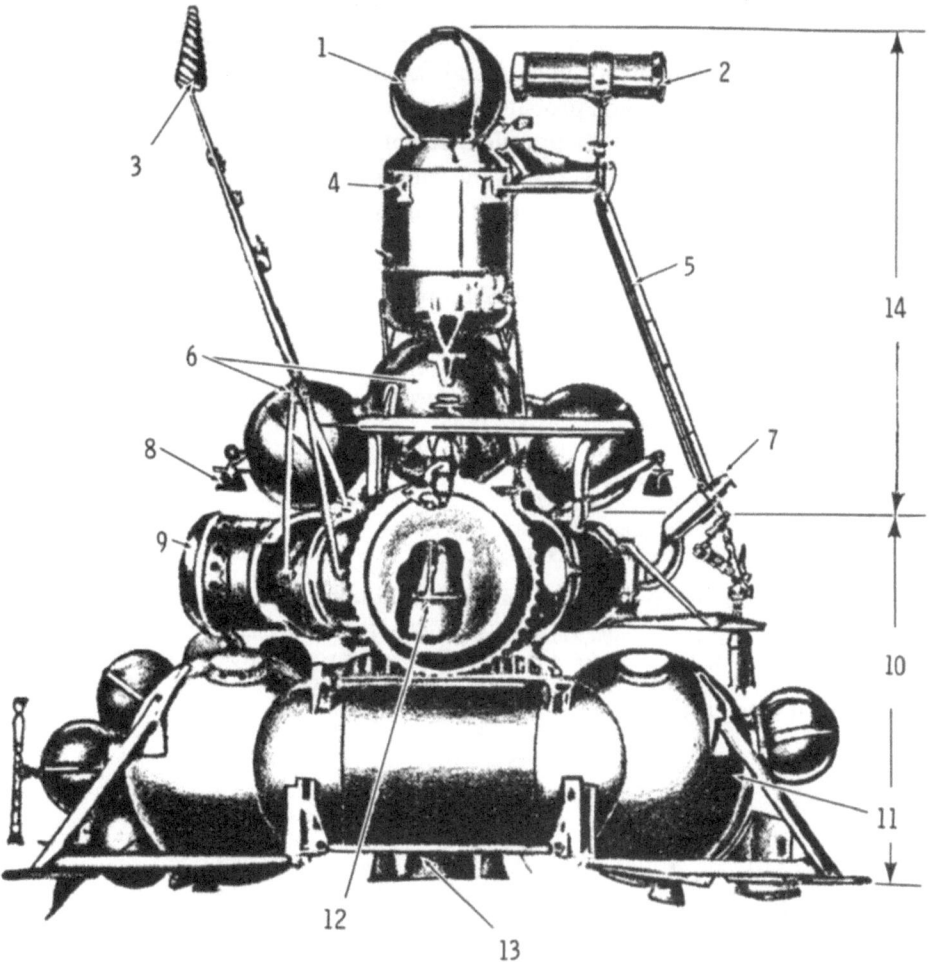

Fig. 2.19.   A schematic view of the U.S.S.R. Luna 16 spacecraft, with equipment for remotely-controlled return of a sample of the lunar soil to the Earth. The successive numbers on the diagram indicate: (1) returning part of the mooncraft; (2) the drilling mechanism; (3) omnidirectional antenna; (4) instrument compartment; (5) drilling mechanism; (6) fuel tanks; (7) camera; (8) attitude vernier nozzle; (9) instrumental compartment of the descent stage; (10) descent stage; (11) fuel for descent; (12) rocket engine for return to the Earth; (13) descent stage engine; and (14) the take-off stage (after Pravda, 1970, No. 227).

By their combined output, all re-entrant spacecraft – manned and unmanned – of American and Russian arigin, described in this chapter, helped to increase our knowledge of the Moon immeasurably. Moreover, no attempt at a comparison of the relative effectiveness of manned and unmanned tools of research in this field by their proven outcome can leave any doubt that – between 1969–1972 – man amply justified his presence on the Moon by having secured for us results which could not have been obtained by any automatic device known at that time. To make only one comparison – the six Apollo missions which landed on the Moon between 1969–1972 brought back to Earth 1000 times as large a mass of lunar rocks (with their highly coded information on the past of our satellite, which will be discussed in Chapters 7 and 8) as did the unmanned landers Luna 16 and 20 within the same time (see Table 7-2 on p. 159). We shall, therefore, probably not err by concluding that, within the 'state of the art' of human technology prevalent around 1970, man in the system was still a necessity for as rapid an advance of our subject as we have seen in the past few years; though whether or not this must remain so in the future opinions still differ.

References

The literature on the Apollo project – at all levels – is truly enormous; so that only its merest outline can be given in this place.

The prime sources of scientific information on the individual missions are as follows:

Apollo 11: *Science* **167**, No. 3918 (pp. 449–784), 1970.
Apollo 12: NASA SP-235, Washington 1970.
Apollo 14: NASA SP-272, Washington 1971.
Apollo 15: NASA SP-289, Washington 1972.
Apollo 16: NASA SP-315, Washington 1972.
Apollo 17: NASA SP-330, Washington 1973.

The following individual references have been quoted in this chapter:

Herschel, William: 1780, see *Scientific Papers of Sir William Herschel*, London 1912, Vol. I, pp. xc–xci.
Kepler, Johannes: 1634, *Somnium sive Astronomia Lunaris*, Francoforti (English translation by John Lear and Patricia Frueh Kirkwood, University of California Press at Los Angeles, 1965).
Nicolson, M. H.: 1948, *Voyages to the Moon*, Macmillan Co., New York.

CHAPTER 3

BASIC FACTS: DISTANCE, SIZE AND MASS

The aim of this chapter will be to introduce to the reader the Moon as a celestial body, and to review its basic properties – such as the distance, size and mass – the knowledge of which will constitute a necessary prerequisite for proper understanding of the structure and other physical characteristics of the Moon discussed later in this book. To the astronomer the Moon has indeed been a friend of old standing; and at least a rudimentary knowledge of its motion goes very far back in the history of mankind on this planet; for since prehistoric times the waxing and waning of lunar phases, and the light changes accompanying them, provided the first astronomical basis for the reckoning of the time. Whenever we go sufficiently far back in the history of almost any primitive civilization, we find them invariably dependent on the lunar, rather than the solar, calendar: the month became a unit of time long before the concept of the year emerged from accumulating observations; and the Moon as the graceful carrier of this knowledge thus gained entrance as a female deity, to the pantheons of many ancient nations.

The measurements of the *distance* which separates us from our nearest celestial neighbour – from ancient times to the advent of the space age – has been reviewed in Chapter 1 of our *Moon II* book, and need not be repeated in this place. Suffice it to say that it varies between 364 400 and 406 730 km each month (due to the eccentricity of the lunar relative orbit). The mean distance of the Moon amounts very closely to 384 400 km, and is equal to 60.268 times the Earth's equatorial radius of 6378.17 km (which would, therefore, be seen from the centre of the Moon at an angle of $\pi = 57'2''.46$ or $3422''.46$, called the Moon's *mean horizontal parallax*), or 0.00257 times the mean distance separating us from the Sun (the latter being, on the average, 389 times as far away from us as the Moon). The distance to the Moon represents, therefore, less than one per cent of the distance separating us from our next two nearest celestial neighbours – the planets Venus and Mars – even at the time of their closest approach. Light traverses this distance in 1.28 s; and a spaceship which can disengage itself from the gravitational field of the Earth can reach the Moon after a flight of 65–70 hr.

The results obtained by astronomical triangulation at a distance were, in more recent past, superseded in accuracy by methods relying on observations of radar or laser echoes sent out from the Earth and returned to the Earth. Only 28 years have elapsed since the first radar contacts with the Moon have been established in 1946 (for fuller details, cf. Chapter 21 of *Moon II*) by groups working independently in Hungary (Bay, 1946) and the United States (DeWitt and Stodola, 1949). Since that time, thousands of radar pulses at different frequencies were beamed on the Moon, and their echoes used to study the structure of the lunar surface (cf. Chapter 9). By accurate timing of the returning echoes (which travel through space with velocity that is sufficiently well known) Yaplee and his collaborators (Burton *et al.*, 1959; Yaplee *et al.*, 1964) re-determined the Earth-Moon distance, and found its mean value to be equal to 384400 ± 1 km – in close accord with previous astronomical determinations.

Still more recently, the Moon has become the target of *laser* beams of coherent light at optical frequencies. The first successful attempt to reflect short-lived flashes of laser light from the lunar surface was made in May 1962 by a group of physicists from the Lincoln Laboratories of the Massachusetts Institute of Technology in U.S.A. (cf. Smullin and Fiocco, 1962). Because of diffuse reflection on the lunar surface, the returning signals were only just strong enough for identification, but not for accurate timing of their return.

This situation has, however, changed drastically ever since the recent spacecraft of American as well as Russian origin – Apollo 11 and 12 as well as Luna 17 and 21 – installed at their different landing sites on the Moon cube-corner reflectors, from which laser signals could be returned with much greater efficiency. With their aid, laser contact with the Moon has been maintained almost continuously from different countries (France, U.S.A., U.S.S.R.); and although few specific results of such work have been published so far, there is no doubt that the laser method is potentially capable of providing instantaneous distances to the Moon with an accuracy of the order of 1–10 m – i.e., 100 to 1000 as high (because of shorter wavelengths employed) as the earlier determinations based on astronomical triangulation or radar ranging.

The *apparent diameter* of the lunar disc in the sky has since ancient times been known to be close to half a degree (Archimedes and Hipparchos adopted for it the round value of '720th part of the zodiac'), and varies somewhat on account of the varying distance of the Moon from the observer on the surface of the Earth as well as on the position in its relative orbit. When the topocentric observations are freed from the parallactic effect and reduced to the centre of the Earth (for the requisite geometry, cf. Chapter 3 of *Moon II*), the mean geocentric apparent diameter of the lunar disc is found to amount to $1865''.2$ – oscillating by $204''.8$ between perigee and apogee – which at the mean distance of 384402 km corresponds to a mean radius of the lunar globe of 1738 km. The Moon is, therefore, little more than one-quarter in size of the Earth. Its surface covers an area of 37.96 million km^2; and its volume amounts to 21.99 milliard km^3 (i.e., approximately 2.03% of that of the Earth).

The next quantity of basic importance for the understanding of the fundamental characteristics of our satellite is its *mass*. The mass of the Moon – like that of any other celestial body – can be determined only from the effects of its attraction on another body of known mass; and in the case of the Moon, this will be our Earth (or, quite recently, an impinging spaceship). There are, in principle, three ways by which this can be done: namely, by a study of the effects of the Moon on (a) the orbital motion of the Earth in space; (b) the axial rotation of the Earth; (c) the infall of a spacecraft.

The cases (a) and (b) have been adequately discussed in Chapter 1 of our previous book (1969), and nothing of significance needs to be added at this time. With incident spacecraft it is, however, a very different story. As this method is based on the measurements (by range-Doppler tracking) of changes in velocity exhibited by a fly-by spacecraft (or impinging hard-lander), the measurements cannot furnish the mass m_{C} of the attracting body (i.e., the Moon) directly, but only to product Gm_{C}, where G stands for the constant of gravitation. The mass-ratios $m_{\oplus}/m_{\text{C}}$ of the Earth:Moon deduced from such measurements in the past ten years are listed in column (2) of the accompanying Table 3-1; while column (3) contains the values of Gm_{C} corresponding to the listed mass-ratios and to the known value of $Gm_{\oplus} = 398\,601 \pm 1$ km^3 s^{-2} for the Earth. The mean value of the ratios $m_{\oplus}/m_{\text{C}}$ listed in Table 3-1 comes out to be equal to 81.302 ± 0.001, and should be correct within one part in 10^5. It corresponds to the value of $Gm_{\text{C}} = 4902.71 \pm 0.06$ km^3 s^{-2}; and, for $G = 6.668 \times 10^{-8}$ cm^3 s^{-1} s^{-2}, to $m_{\text{C}} = 7.353 \times 10^{25}$ g as the best approximation we possess to the actual mass of the Moon (with an uncertainty due solely to our incomplete knowledge of the gravitation constant G).

The lunar mass of 7.353×10^{25} g or over 73 trillion tons may loom large in comparison with all terrestrial standards; but on cosmic scales it constitutes but a relatively tiny speck. And neither is the mean *density* of the lunar globe at all unusual; for dividing the lunar mass just found by the volume of 2.199×10^{25} cm^3, we obtain its mean density to be $\varrho = 3.34$ g cm^{-3} – i.e., about the same as that of common basaltic rocks of the Earth's crust, and considerably smaller than the mean density of the terrestrial globe (5.53 g cm^{-3}). The mean gravitational acceleration $Gm_{\text{C}}/r_{\text{C}}^2$ on the lunar surface is, therefore, only 162 cm s^{-2} (i.e., less than one-sixth of the terrestrial one); and the mean *velocity of escape* $(2Gm_{\text{C}}/r_{\text{C}})^{1/2}$ from the lunar gravitational field is close to 2.38 km s^{-1} – in comparison with the terrestrial value of 11.2 km s^{-1}.

TABLE 3-1

Values of the Earth-Moon mass-ratio and of the mass of the Moon deduced from spacecraft motions

Spacecraft	Mass-ratio $m_{\oplus}/m_{\text{C}}$	Gm_{C}(in km^3 s^{-2})
Mariner 2 (1962)	81.3001 ± 0.0013 (m.e.)	4902.83 ± 0.08
Mariner 4 (1965)	81.3015 ± 0.0017	4902.75 ± 0.11
Mariner 5 (1967)	81.3006 ± 0.0008	4902.80 ± 0.06
Mariner 6 (1969)	81.3011 ± 0.0015	4902.77 ± 0.10
Mariner 7 (1969)	81.2997 ± 0.0015	4902.86 ± 0.10
Pioneer 6 (1965)	81.3005 ± 0.0007	4902.81 ± 0.06
Pioneer 7 (1966)	81.3021 ± 0.0004	4902.72 ± 0.03
Ranger 6 (1964)	81.3930 ± 0.0030	4902.66 ± 0.14
Ranger 7 (1964)	81.3050 ± 0.0027	4902.54 ± 0.15
Ranger 8 (1965)	81.3035 ± 0.0020	4902.63 ± 0.12
Ranger 9 (1965)	81.3022 ± 0.0047	4902.71 ± 0.20
Lunar Orbiters (1966–1968)	81.3033 ± 0.0018	4902.64 ± 0.11

The *actual* velocity of escape from the gravitational field of the Moon – of importance for problems connected with the escape of gases from the lunar exosphere (cf. Chapter 9) will, of course, be modified somewhat by the attraction of the Earth (and the Sun). In order to investigate quantitatively this particular aspect of the problem, let in what follows $m_\oplus$, m_ζ continue to stand for the masses of the Earth and the Moon, respectively; and let m represent a particle of negligible mass (such as an atom of gas molecule) moving in a gravitational dipole Earth-Moon, referred to a system of rectangular coordinates XYZ, with origin at the centre of mass of the Earth-Moon system, and rotating in space with the system about the Z-axis. If so, the well-known Jacobian energy integral of the corresponding restricted problem of three bodies assumes the form

$$C = \frac{2(1-\mu)}{\varDelta} + \frac{2\mu}{r} + X^2 + Y^2 - V^2,$$

(3.1)

where C denotes a constant of integration; V, the velocity of the mass-particle m in the frame of reference of our rotating gravitational dipole;

$$\mu = \frac{m_\zeta}{m_\oplus + m_\zeta} = 0.01215$$

(3.2)

and $\varDelta$, r denote the distances of m from $m_\oplus$ and m_ζ, respectively. If the mean distance $R = 384400$ km separating $m_\oplus$ and m_ζ is hereafter adopted as our unit of length, and the sum $m_\oplus + m_\zeta$ as that of mass, the corresponding units of time t and velocity V become $[G(m_\oplus + m_\zeta) R^{-3}]^{-1/2} = 4.3423$ days and $[G(m_\oplus + m_\zeta) R^{-1}]^{1/2} = 1.0246$ km s^{-1}, respectively.

For a particle situated on the lunar surface of mean radius $r_\zeta = 1738$ km,

$$r = \frac{r_\zeta}{R} = \frac{1738}{384400} = 0.004\,521 ;$$

(3.3)

while X varies between $1 - \mu \pm (r_\zeta/R)$ – so that

$$0.9834 \leqslant X \leqslant 0.9924,$$

and

$$Y \leqslant \tan(r_\zeta/R) = 0.004\,52.$$

Accordingly,

$$X^2 + Y^2 \simeq 0.9759$$

(3.4)

and

$$\varDelta \simeq \sqrt{1 - (r_\zeta/R)^2} = 0.999\,989\,8 ;$$

(3.5)

in consequence of which Equation (3.1) assumes the particular form

$$C = 8.3261 - V^2.$$

(3.6)

Now a geometry of the surfaces defined by Equation (3.1) reveals (cf., e.g., Kopal, 1959, Chapter III) that, for $\mu = 0.01215$, the following five values of C:

$$C_1 = 3.1876,$$
$$C_2 = 3.1707,$$
$$C_3 = 3.0122, \tag{3.7}$$
$$C_{4,5} = 2.9878,$$

are of critical significance. For $C \geqslant C_1$ Equation (3.1) represents two oval surfaces closed around the two finite mass centres. For $C = C_1$ these two ovals come in contact and leave the Moon to fall towards the Earth if endowed with a velocity as low as 2.2 km s^{-1}; and only a slightly higher velocity could enable them to leave the Earth-Moon system altogether and to be captured by the Sun. Any particles ejected from the lunar surface with velocities less than 2.2 km s^{-1} but more than approximately $\sqrt{Gm./r} = 1.68$ km should, however, orbit around the Moon within the space limited by the close oval surrounding the Moon for $C = C_1$ – the mean radius of which is approximately 38000 km (i.e., more than twenty times the radius of the Moon itself). Only if $V^2 < Gm./r$, will the particle so ejected fall back on the lunar surface along trajectories which have been discussed in Chapter 17 of *Moon II*.

We may add that even the foregoing formulation of the problem of gravitational escape from the Moon is not yet exact, as we disregarded the finite eccentricity of the relative orbit of the Earth-Moon gravitational dipole. When this eccentricity (which oscillates between 0.0432 and 0.0666 around its mean value of 0.0549) is taken into account, not only R, but also C become functions of the time (cf. Kopal and Lyttleton, 1963; or *Moon II*, Chapter 6); and their fluctuations must be considered in any predictions of the actual velocities of escape at any particular time.

References

Bay, Z.: 1946, *Hungarian Acta Phys.* 1, 1.

Burton, R. H., Craig, K. J., and Yaplee, B. S.: 1959, *Astron. J.* **64**, 325.

DeWitt, J. H. and Stodola, E. K.: 1949, *Proc. I.R.E.* **37**, 229.

Kopal, Z.: 1959, *Close Binary Systems* (Chapman-Hall and John Wiley, London and New York), Chapter III.

Kopal, Z. and Lyttleton, R. A.: 1963, *Icarus* 1, 455.

Smullin, L. D. and Fiocco, G.: 1962, *Nature* **194**, 1267.

Yaplee, B. S., Knowles, S. H., Shapiro, A., Craig, K. J., and Brouwer, D.: 1964, US Naval Research Laboratory Rept. No. 6134.

CHAPTER 4

SHAPE AND GRAVITATIONAL FIELD OF THE MOON

That the shape of the Moon is very approximately spherical is a fact which can be verified with the naked eye whenever the Moon happens to be on the horizon; and in the preceding chapter we outlined the methods by which its mean radius was established at very close to 1738 km. Moreover, in the next chapter we shall explain why a self-gravitating body of the mass of the Moon was bound to have settled to this form. The aim of this chapter will be to review the extent to which the actual shape of the Moon deviates from a sphere of radius equal to 1738 km, and its gravitational field departs from that of a globe with spherically-symmetrical distribution of matter. These deviations – while small – have an important message to deliver concerning the state of equilibrium of the lunar globe, and its internal structure. An inquiry into this structure will be conducted in the next chapter on the basis of all aspects of observational evidence bearing on the problem; and in the present chapter we shall begin to prepare the ground for such an inquiry.

While direct observations have long disclosed the Moon to be very approximately a sphere, deviations from spherical shape began to manifest themselves only since the increasing precision of observations enabled us to establish global deformations of the order of 0.1% of the Moon's radius. There are, in principle, two classical methods by which the detailed shape of the Moon can be studied at a distance from the Earth. The most readily available information is provided by the shape of the Moon's marginal zone – i.e., of the region which appears in projection on the celestial sphere as the limb of the apparent lunar disc. The optical librations of our satellite (cf. Chapter 3 of *Moon II*) will bring from time to time almost 17.7% of the entire lunar surface to the limb of apparent disc in the sky; and the shape of this limb can then be measured by the usual astrometric methods. For fuller details of the data obtained by these methods cf. Kopal and Carder (1974), with references quoted therein.

The second method is based on observations of the angular displacements of individual points on the lunar surface with respect to the limb (or to an absolute frame of reference as represented, for instance, by the background stars), and takes advantage of the fact that features on a warped surface are displaced for the terrestrial observer in a different way at different libration angles than they would if they were located on a sphere. For details of this stereoscopic method cf. Chapter IV of Kopal and Carder (1974). The method is no longer limited to limb regions, but can be applied to any part of the visible lunar surface. Its principal limitation is, however, the smallness of parallactic shifts of lunar features on a warped surface caused by lunar librations. The Moon is so far from us, and the amplitude of lunar librations so small, that astrometric observations of highest accuracy are required to furnish hypsometric data of any significance – beyond a disclosure that departures of the lunar surface from a sphere lie mostly within ±3 km in absolute elevation, and are geometrically quite complicated.

In particular – unlike the Earth – the Moon does not exhibit any noticeable flattening at the poles of axial rotation (a fact which should cause no surprise, as the equatorial velocity of rotation of so small a globe with the period of one month should cause a difference of only 16 m between the equatorial and polar semi-axis); nor any noticeable elongation in the direction of the Earth; and any notion of a 'tidal bulge' has to be relegated to the domain of early folklore of the subject. The actual figure of the Moon, as it began to emerge from intensive work of the last decade (cf. Meyer and Ruffin, 1965; Eigen and Hathaway, 1967; Mills, 1967, 1968; Mills and Sudbury, 1968), appeared to be so complicated that no spheroid or three-axial ellipsoid (of any arbitrary orientation of its axes) provided any better approximation than a sphere.

When we exchange Earth-based for space-borne photography, improved results should be expected on account of a greater proximity to the target, as well as of a complete freedom from the deleterious atmospheric effects on the quality of photographic images. The usefulness of the Lunar Orbiter photographs of 1966–1967 for hypsometric purposes was limited by the fact that only second-generation photographic data (transmitted by readout) were available to us on the Earth. This situation changed drastically in our favour with the advent of re-entrant missions; and, in particular, since the commencement of orbital science with Apollo 15 (cf. Chapter II). The Apollo 15 Control System, prepared recently by the U.S. Defense Mapping Agency Aerospace Center for NASA (cf., Helmering, 1973; Schimerman et al., 1973, and others) has established some 4900 equally spaced spacecraft-based positions and absolute elevations over approximately 10% of the lunar surface within sight of the mission's metric cameras. Apollo 16's data remain so far unreduced; and as Apollo 17's overflight track overlapped very largely with that of Apollo 15, not much new can be expected from it in this respect.

Until the advent of lunar spacecraft, stereoscopic effects exhibited in the course of lunar librations and observable from the Earth offered the only way to find out anything about the departures of our satellite from spherical shape. The first contributions of novel kind made by spacecraft came from the hard-landing Rangers of 1964–1965. Determinations of the positions in space of points at which the free-flight trajectories of these spacecraft came to their end on crash-landing led to the data summarized in Table 4-1 (after Sjogren, 1967); and furnished four absolute radii of the Moon, connecting the landing points with the centre of mass of our satellite, accurate within ±100 m.

The outstanding feature of the foregoing results was the realization that the front side of the Moon – far from being elongated in the direction of the Earth – is actually depressed below the mean moon-level by some 2.6 km; and this important result has been amply confirmed since that time.

TABLE 4-1

Absolute lunar radii inferred from the impacts of Ranger spacecraft

Spacecraft	Data and time of impact	Coordinates of impact point		Local radius (in km)
Ranger 6	1964 Feb 2, $9^h24^m33^s$ UT	21°52 W	9°33 N	1735.3
Ranger 7	1964 July 31, 13 25 49	20.58 E	10.63 S	1735.5
Ranger 8	1965 Feb 20, 9 57 37	24.64 W	2.67 N	1735.2
Ranger 9	1965 Mar 24, 14 7 20	2.37 E	12.83 S	1735.7

Hypsometric results of much greater significance were obtained since the advent of the Apollo project in 1969. As is well known, Apollo 11 and 12 missions, in 1969, followed by the Luna 17 in 1970, deposited on the lunar surface cube-corner reflectors, capable of returning laser signals flashed from the Earth; and from the timing of the return of such light 'echoes' from the Moon the instantaneous distance between the transmitter on Earth and the reflector on the Moon can be determined within one part in 10^8 (i.e., ± 4 m). Since this distance varies with the time as a result of the continuous relative motion of both stations, long series of observations are needed to specify the absolute elevation of each respective cube-corner. The requisite measurements have been in progress for some time; and although few specific results have been reported so far (cf. Williams *et al.*, 1973), more can be confidently expected in the future.

However, Apollo 15 and 16 missions of 1971–1972 made much more extensive contributions to our knowledge of the shape of the Moon by means of the laser altimeters mounted in the scientific instruments module of the Command and Service modules of the respective missions, which were revolving in nearly circular orbits around the Moon while the Excursion module descended to the surface. These orbits were monitored continuously by the tracking stations on the Earth. The position of the spacecraft relative to the Moon's centre thus being known with a few metres, and its altitude above the actual surface being determined from the time-delay of returning laser echoes with the same accuracy, the difference between the two will furnish the distance of the reflecting element of the surface from the Moon's centre within total errors not exceeding ± 10 m (i.e., 10 to 100 times smaller than any data previously obtainable from the distance of the Earth).

The altitude profiles of the Moon measured by the laser altimeters of the Apollo 15 and 16 missions (cf. Wollenhaupt and Sjogren, 1972) are shown on the accompanying Figures 4.1 and 4.2. Each set of these data refers to a different cross-section of the orbit of the respective spacecraft with the surface of the Moon; and cannot say directly anything on absolute elevations anywhere else. But they disclose that – along each cross-section – the actual lunar surface deviates from the mean sphere of 1738 km radius in a very complicated manner, not describable by any single (or a small number of) surface harmonics. The outlines of individual maria or mountain chains stand out clearly on the continental background; and other local features known from surface photography can be distinguished in the records.

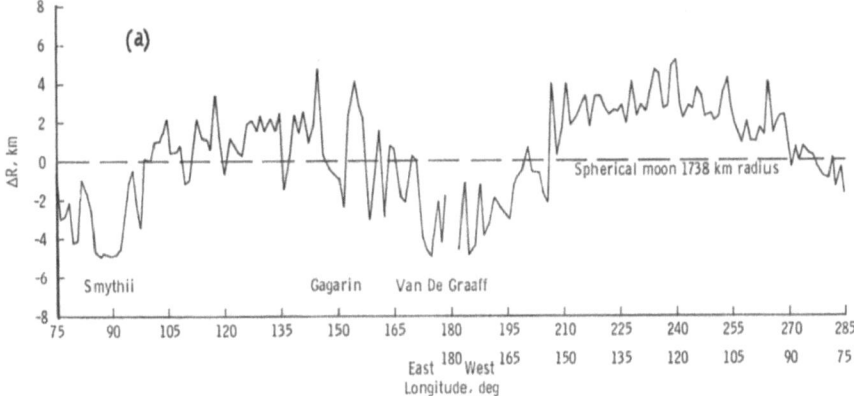

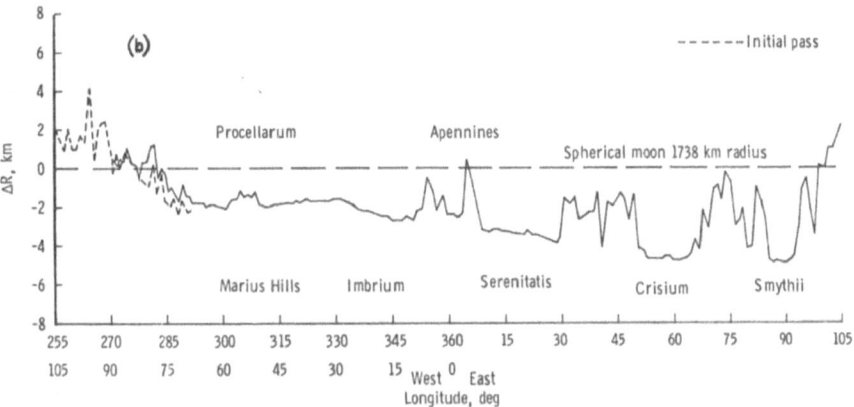

Fig. 4.1. Altitude profile and linear deviations from a spherical Moon of the mean radius of 1738.0 km, as measured by the Apollo 15 laser altimeter along the entire ground-track of its traverse in July 1971: (a) lunar far side; (b) lunar near side (after W. R. Wollenhaupt and W. L. Sjogren, *Moon* **4**, 337, 1972).

Perhaps the most important result furnished by the laser altimeters of Apollo 15 and 16 has been the confirmation that the front hemisphere of the Moon – far from being elongated in the direction of the Earth in any kind of 'fossil tide' – is actually depressed below the mean moon-level; and it is the far side which appears to be elongated. From this fact one could conclude that the centre of symmetry of the actual figure of the Moon is displaced by 2–3 km from the Moon's centre of mass. Such an interpretation of the observed facts – while possible – is, however, not yet necessary. Any asymmetry between the front and far sides of the Moon, as seen from the Earth, is describable in terms of the odd harmonics in selenographic longitude, into which any particular cross-section can be decomposed. A presence, in such an expansion, of the first harmonic $(j=1)$ would necessitate indeed the centre of figure to be displaced from the centre of mass, to an extent consistent with its amplitude. If, however, the observed asymmetry of figure can be represented by a combination of harmonics higher than the first (corresponding to $j=3$, 5, etc.) with appropriate coefficients, no shift between centre of figure and of mass is required any more to account for the observed facts. The two altitude profiles furnished by Apollo 15 and 16 are insufficient to decide without ambiguity which of the two alternatives may represent better the observed facts; and as no more Apollo flights are scheduled to contribute further to the available evidence, the issue is likely to remain undecided for some time to come.

Another important result which transpired from the Apollo 15–16 laser altimetry was to dispel another previously cherished notion even more firmly entrenched in lunar folklore than the 'fossil tide': namely, a contention that all continental areas on the Moon (i.e., regions of relatively high albedo) are 'highlands'. It is true (cf. again Figures 4.1 and 4.2) that the plains of several maria seen on the front side and over-flown by Apollos 15–16 are appreciably depressed below the level of the surrounding landscape. But, on the other hand, typically continental regions were found (especi-ally on the Moon's far side) which lie even lower (i.e., nearer to the centre of the Moon) than the maria, but are not filled with any mare basalts. This is, in particular, true of a great trench of low-lying ground running approximately north-south between

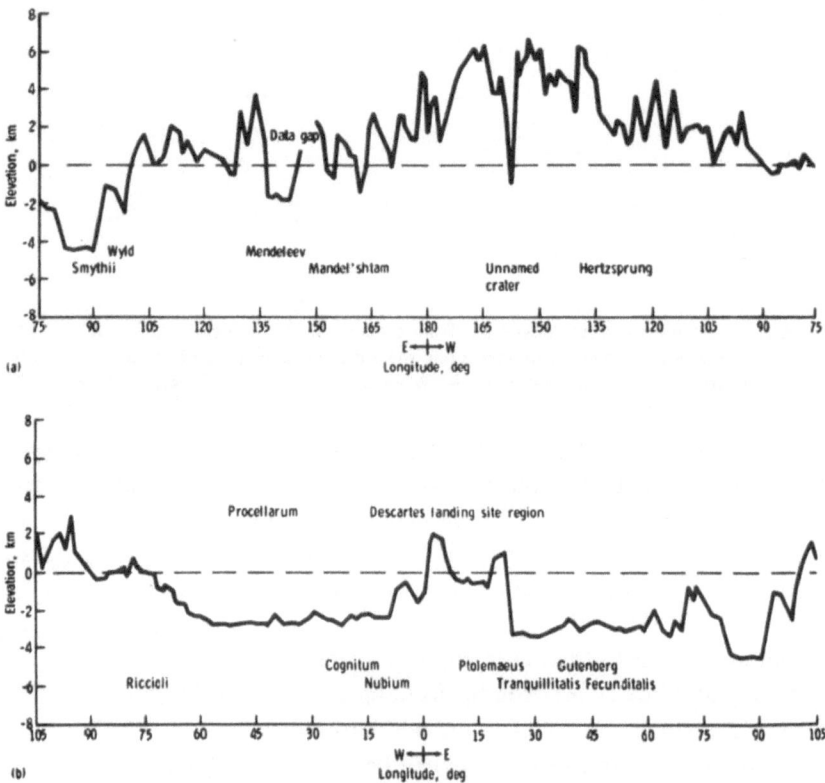

Fig. 4.2. Apollo 16 altitude profile, showing the deviations of the actual lunar surface from a mean sphere of 1738.0 km radius: (a) lunar far side, (b) lunar near side (after Wollenhaupt and Sjogren, *Apollo 16 Prelim. Sci. Rept.*, NASA SP-315, sec. 30A, 1972).

170°–180° of eastern longitude on the Moon's far side, discovered by the Soviet spacecraft Zond 6 in 1968 (cf. Rodionov *et al.*, 1971), and confirmed by the Apollo 15 laser altimeter three years later.

The implications of all these data for the internal structure and evolution of the Moon will be discussed more fully in the subsequent Chapters 5 and 10. For the present, we wish to supplement them with parallel information which we have accumulated on the *gravitational field* of our satellite.

If the Moon were an exact sphere, with a spherically-symmetrical distribution of mass in its interior, its external gravitational field should be that of a point of mass equal to that of the Moon as a whole. The first – and sufficient – indication that the Moon does not conform to so simple a model came to us from the familiar phenomenon of synchronism between rotation and revolution of our satellite – in brief, from the fact that the Moon continues to show us the same face. If its gravitational potential were that of a mass-point, the Earth's attraction would be powerless to effect any such synchronization; and the Moon could rotate as it wished about any arbitrary axis.

Since this is not the case, the moments of inertia of the lunar globe about different axes must obviously differ; and the extent of their differences was disclosed to us first by the 'physical librations' of our satellite – as distinct from the 'optical librations', to which we referred in the first part of this chapter. It is the latter that enable us to see a little 'over the limb' of a particular hemisphere by virtue of the eccentricity of the relative lunar orbit (which makes a difference of the angular velocity of rotation and revolution fluctuate somewhat in the course of each month) and of its deviation (by $5°9' \pm 1°32'$) from normality to the axis of rotation. In addition to such librations (depending on the characteristics of lunar orbit) our satellite exhibits, however, also 'physical librations' which represent the motions of the Moon as a rigid body, about its centre of mass, caused by the attraction of the Earth as well as of the Sun. A theory of such librations constitutes a problem of some complexity, which we discussed already in Chapter IV of *Moon II*; and its details need not be reproduced in this place. Suffice it to say that – unlike the optical librations which may attain the magnitude of several degrees – the selenocentric amplitudes of the physical librations of the Moon do not exceed two minutes of arc. At the distance of the Earth, such librations do not cause displacements exceeding $0''.54$ – a fact which demonstrates the difficulty of measuring them with any significance.

It was one of the unheralded triumphs of positional astrometry in the century between 1840–1950 to have actually measured such displacement against odds which are difficult to explain to the uninitiated; and to have extracted from the periods of such motions the ratios

$$\frac{C-B}{A}=\alpha, \qquad \frac{C-A}{B}=\beta, \qquad \frac{B-A}{C}=\gamma, \tag{4.1}$$

of the moments A, B, C about the principal axes of inertia of the lunar globe; satisfying (by virtue of their definition) the equation

$$\alpha - \beta + \gamma = \alpha\beta\gamma,\tag{4.2}$$

which can be solved for one of these three quantities if the other two are known.

In order to determine β, all we need to establish from the observations in principle are the angles i and I of inclination of the Moon's orbit and equator to the ecliptic, and the rate m of regression of lunar nodes expressed in terms of the Moon's angular velocity of revolution; for, to a good approximation (cf. *Moon II*, Chapter 4),

$$\beta = \tfrac{2}{3}\left(\frac{mI}{i+I}\right).\tag{4.3}$$

Now the angle i is known very accurately from the Moon's motion to be equal to 5°8′43″.4 (Brown, 1919); while $I = 1°32'4''$ (Koziel, 1967a, b). Moreover, the period of regression of the nodes of lunar orbit is known again with great precision to equal 18.6133 tropical years or 248.827 sidereal months, rendering $m = (248.827)^{-1} = 0.00401886$. If so, however, Equation (4.3) yields $\beta = 0.000615$; and its refinement taking account of the squares of orbital eccentricity e and inclination i led to

$$\beta = 0.000628 \pm 0.000001\tag{4.4}$$

whose uncertainty stems mainly from that of the observed value of I.*

While the value of β can, in effect, be specified in terms of the amplitudes of lunar physical librations, that of γ can be deduced from their periodicity. The arduous task of extracting it from heliometric observations extending over almost one century was recently carried out by Koziel (1967a, b), and led to a value of

$$\gamma = 0.000230 \pm 0.000006;\tag{4.5}$$

while the value of α, corresponding to those of β and γ as given above, resulted from the identity (4.2) as

$$\alpha = 0.000397 \pm 0.000008.\tag{4.6}$$

The foregoing values of α, β, γ as given by Equations (4.4)–(4.6) are based solely on ground-based heliometric observations. They may be compared with the values

$$\alpha = 0.000405 \pm 0.000003,$$
$$\beta = 0.000636 \pm 0.000005,\tag{4.7}$$
$$\gamma = 0.000226 \pm 0.000003,$$

inferred more recently from laser tracking (cf. Bender *et al.*, 1973). The internal agreement between the two sets of data – based on entirely different observational approach – is excellent, and inspires full confidence in their correctness.

* Recent results of the Apollo 15 mission (cf. Wollenhaupt *et al.*, 1972) indicated that – for the epoch of 1971. 6 – the actual value of I was by 135″ larger than derived by Koziel, which referred to a mean equinox close to 1900. Whether or not the difference between the two reflects a secular increase in I (corresponding to a change of approximately $+2''$ per annum) is not yet certain, but appears a distinct possibility (cf. Kopal, 1972).

The extent to which the foregoing values of α, β and γ differ from zero constitutes a testimony on the departure of the Moon from spherical form. On their own, they are insufficient to define the actual shape of our satellite – but such form (and internal structure) as the Moon may possess must be consistent with them. The fact that they differ from zero constitutes, however, a sufficient cause to enable the Earth to get the gravitational 'handle' on the Moon and to synchronize, in time, its rotation with the revolution.

A further discussion of the full significance of the actual values of α, β, γ as physical constraints on the present structure of the Moon and its past evolution is being deferred to subsequent Chapters 5 and 10; at present we wish to continue with an account of our present knowledge of the more detailed 'anatomy' of the Moon's gravitational field. Until 1966, physical librations represented the only way in which we could probe certain properties of the gravitational field of our satellite. Artificial satellites of the Moon, launched that year in considerable numbers (cf. Tables 1-1d and 1-3), could – in addition to their photographic missions – be employed also as sensors of the lunar gravitational field, in which they were navigating in free flight for months at a time. Accurate Doppler tracking of the radio signals transmitted by such satellites (but limited, alas, to only one-half of the orbit, when the spacecraft is not occulted by the Moon) can be used for precise reconstruction of their space motions; and, from the perturbations of their elements, of the Moon's external gravitational field.

In more specific terms, let the external gravitational potential V of the lunar globe be expressed in the form of an expansion

$$V = \frac{Gm_{\mathfrak{c}}}{r}\left\{ C_{0,0} + \sum_{n=2}^{\infty}\sum_{m=0}^{n} \left(\frac{r_{\mathfrak{c}}}{r}\right)^n P_n^m(\sin\beta)\left[C_{n,m}\cos m\lambda + S_{n,m}\sin m\lambda\right]\right\},$$

$$(4.8)$$

where β, λ denote the selenographic latitude and longitude; r, the distance of the spacecraft from the centre of mass of the Moon; P_n^m, the associated Legendre coefficients of order m and degree n; and $C_{n,m}$, $S_{n,m}$ the coefficients specifying the gravitational field of the Moon. A recent analysis of some 20 000 individual observations between 1966–1968 by Michael and Blackshear (1972), carried out to terms of 13th degree, led to the results summarized in Table 4-2; and a map of the corresponding gravimetric anomalies is shown on Figure 4.3.

The value given for $C_{0,0}$ corresponds to $Gm_{\mathfrak{c}} = 4902.84$ km^3 s^{-2} – in good agreement with the value of 4902.71 deduced in the preceding chapter from the observed accelerations of fly-by and hard-landing mooncraft; and the mean radius $r_{\mathfrak{c}} = 1738.09$ km came out likewise very close to the one which we adopted so far.

The second-degree coefficients C_2^m and S_2^m of the lunar gravitational potential can be related with the moments A, B, C of inertia of the Moon by the equations

$$C_{2,0} = \frac{A+B-2C}{2m_{\mathfrak{c}}r_{\mathfrak{c}}^2}$$

$$(4.9)$$

THE MOON IN THE POST-APOLLO ERA

TABLE 4-2

Coefficients of the thirteenth degree and order solution for the lunar gravitational field*

n	m	$C_{n,m}$	$S_{n,m}$
0	0	1.00005388274	
2	0	$-2.0378761820 \times 10^{-4}$	0
	1	$1.1051542512 \times 10^{-5}$	$1.3006148287 \times 10^{-5}$
	2	$2.4845211490 \times 10^{-5}$	$-1.0442661244 \times 10^{-8}$
3	0	$2.8439912986 \times 10^{-5}$	0
	1	$2.4152698939 \times 10^{-5}$	$2.0805931299 \times 10^{-5}$
	2	$7.6322652451 \times 10^{-6}$	$2.2710804784 \times 10^{-6}$
	3	$1.4111967268 \times 10^{-6}$	$-3.1126414977 \times 10^{-7}$
4	0	$3.4688330254 \times 10^{-5}$	0
	1	$-1.9891933273 \times 10^{-5}$	$-8.5100449545 \times 10^{-6}$
	2	$-2.5443614074 \times 10^{-6}$	$-4.1656170604 \times 10^{-6}$
	3	$5.6321036284 \times 10^{-7}$	$2.7682641529 \times 10^{-7}$
	4	$-5.3981736976 \times 10^{-8}$	$1.1411256460 \times 10^{-7}$
5	0	$2.6623483022 \times 10^{-5}$	0
	1	$-7.4939720619 \times 10^{-6}$	$-0.253582656 76 \times 10^{-6}$
	2	$3.4603506756 \times 10^{-7}$	$3.8762399793 \times 10^{-7}$
	3	$4.0040015934 \times 10^{-7}$	$8.8773988822 \times 10^{-7}$
	4	$9.3924925003 \times 10^{-8}$	$-1.4794482739 \times 10^{-7}$
	5	$1.8771823966 \times 10^{-8}$	$2.1001366939 \times 10^{-8}$
6	0	$4.9673196526 \times 10^{-5}$	0
	1	$-1.8543318257 \times 10^{-5}$	$-9.5673010658 \times 10^{-6}$
	2	$-5.5091031933 \times 10^{-7}$	$1.8138036159 \times 10^{-7}$
	3	$1.1678358571 \times 10^{-7}$	$1.1841174024 \times 10^{-7}$
	4	$-3.3956313198 \times 10^{-8}$	$-9.7021389127 \times 10^{-8}$
	5	$-9.6847691920 \times 10^{-9}$	$8.1237744599 \times 10^{-9}$
	6	$-4.0790023093 \times 10^{-9}$	$2.9739566973 \times 10^{-9}$
7	0	$+7.2781850970 \times 10^{-5}$	0
	1	$4.1810263527 \times 10^{-6}$	$2.8408093790 \times 10^{-5}$
	2	$-8.3378247684 \times 10^{-7}$	$7.4166186122 \times 10^{-7}$
	3	$-1.2737535124 \times 10^{-7}$	$-1.7509754956 \times 10^{-7}$
	4	$-4.1682647653 \times 10^{-8}$	$-2.5407115475 \times 10^{-8}$
	5	$1.8658206027 \times 10^{-9}$	$1.0910340047 \times 10^{-8}$
	6	$4.6261404574 \times 10^{-10}$	$-7.1046094393 \times 10^{-10}$
	7	$3.4334519153 \times 10^{-10}$	$-1.2401781124 \times 10^{-10}$
8	0	$1.7411742627 \times 10^{-5}$	0
	1	$1.3701858067 \times 10^{-5}$	$3.6716642133 \times 10^{-6}$
	2	$4.0795796661 \times 10^{-7}$	$9.1385019071 \times 10^{-7}$
	3	$9.3910161964 \times 10^{-9}$	$-1.9359154697 \times 10^{-7}$
	4	$-6.4523074373 \times 10^{-9}$	$2.5059489581 \times 10^{-8}$
	5	$4.5024517927 \times 10^{-9}$	$-9.1723952804 \times 10^{-10}$
	6	$3.6434678341 \times 10^{-10}$	$1.0430917747 \times 10^{-9}$
	7	$-1.4858612477 \times 10^{-11}$	$2.1525564980 \times 10^{-11}$
	8	$-3.6012675699 \times 10^{-11}$	$1.1935027479 \times 10^{-11}$

* After Michael and Blackshear (1972).

Table 4-2 (Continued)

n	m	$C_{n,m}$	$S_{n,m}$
9	0	$-6.5101326253 \times 10^{-5}$	0
	1	$5.3770804532 \times 10^{-7}$	$-5.2273480368 \times 10^{-6}$
	2	$-4.4206541496 \times 10^{-7}$	$-3.1727246486 \times 10^{-7}$
	3	$-1.2910912315 \times 10^{-7}$	$-5.4900413900 \times 10^{-8}$
	4	$8.3591988861 \times 10^{-9}$	$2.1240873880 \times 10^{-8}$
	5	$8.7259392768 \times 10^{-10}$	$-2.5124288158 \times 10^{-9}$
	6	$-5.1505398116 \times 10^{-10}$	$1.2516041989 \times 10^{-10}$
	7	$-8.0690242529 \times 10^{-11}$	$8.0754026764 \times 10^{-11}$
	8	$-1.2649479578 \times 10^{-12}$	$2.6681227345 \times 10^{-11}$
	9	$1.7805101055 \times 10^{-12}$	$-5.4376209997 \times 10^{-13}$
10	0	$-1.9534030711 \times 10^{-5}$	0
	1	$3.5640472538 \times 10^{-5}$	$4.51233094077 \times 10^{-7}$
	2	$4.9196310871 \times 10^{-9}$	$5.7169685637 \times 10^{-7}$
	3	$-2.2358784038 \times 10^{-9}$	$-3.6437585575 \times 10^{-8}$
	4	$3.9141653049 \times 10^{-9}$	$1.1269567323 \times 10^{-8}$
	5	$3.8087666364 \times 10^{-10}$	$-2.1308938653 \times 10^{-9}$
	6	$-4.6255135189 \times 10^{-11}$	$1.5652458548 \times 10^{-10}$
	7	$2.7614962503 \times 10^{-11}$	$-1.5540376682 \times 10^{-11}$
	8	$5.4896743830 \times 10^{-12}$	$-4.0505042040 \times 10^{-12}$
	9	$9.2407535252 \times 10^{-14}$	$-3.6136326596 \times 10^{-13}$
	10	$-7.5223488683 \times 10^{-14}$	$3.3135927064 \times 10^{-14}$
11	0	$5.1037452530 \times 10^{-5}$	0
	1	$-7.7291781814 \times 10^{-6}$	$2.8479446434 \times 10^{-5}$
	2	$-3.2273658086 \times 10^{-7}$	$7.2497774428 \times 10^{-7}$
	3	$2.3360853505 \times 10^{-8}$	$5.5966682329 \times 10^{-8}$
	4	$1.0287849331 \times 10^{-9}$	$1.3205804852 \times 10^{-8}$
	5	$-3.1877004821 \times 10^{-10}$	$-8.8603014695 \times 10^{-10}$
	6	$6.2535668259 \times 10^{-11}$	$1.2557575513 \times 10^{-10}$
	7	$-1.3948668302 \times 10^{-12}$	$-1.6191981587 \times 10^{-11}$
	8	$-9.6800165098 \times 10^{-13}$	$2.8512390282 \times 10^{-13}$
	9	$-2.8524797963 \times 10^{-13}$	$2.5679558039 \times 10^{-13}$
	10	$1.4876779292 \times 10^{-14}$	$2.0466569257 \times 10^{-14}$
	11	$7.2963987090 \times 10^{-15}$	$-1.1859856626 \times 10^{-15}$
12	0	$9.7617164640 \times 10^{-6}$	0
	1	$1.2778274406 \times 10^{-5}$	$1.4438415728 \times 10^{-6}$
	2	$-1.5045505424 \times 10^{-7}$	$5.3551638732 \times 10^{-7}$
	3	$5.4002051640 \times 10^{-9}$	$8.7868585746 \times 10^{-9}$
	4	$4.0078293255 \times 10^{-9}$	$1.2737288951 \times 10^{-9}$
	5	$9.5030429269 \times 10^{-11}$	$-4.8376477908 \times 10^{-10}$
	6	$1.7093579633 \times 10^{-11}$	$6.0134093338 \times 10^{-12}$
	7	$-7.2263198820 \times 10^{-13}$	$-5.5677059494 \times 10^{-12}$
	8	$1.2421199266 \times 10^{-13}$	$9.7510387449 \times 10^{-13}$
	9	$-3.7285776082 \times 10^{-14}$	$-2.4368928834 \times 10^{-14}$
	10	$9.9312843321 \times 10^{-15}$	$-1.0618795919 \times 10^{-14}$
	11	$5.2033681565 \times 10^{-16}$	$-4.2870250967 \times 10^{-16}$
	12	$-2.3855459682 \times 10^{-18}$	$4.5543340194 \times 10^{-18}$
13	0	$7.5782274244 \times 10^{-6}$	0
	1	$-5.7637379713 \times 10^{-6}$	$2.3610613996 \times 10^{-5}$
	2	$1.5991666946 \times 10^{-7}$	$3.4297075982 \times 10^{-7}$

Table 4-2 (Continued)

n	m	$C_{n,m}$	$S_{n,m}$
	3	$2.7190638195 \times 10^{-8}$	$9.6757608497 \times 10^{-9}$
	4	$7.6732684554 \times 10^{-10}$	$3.0028912747 \times 10^{-9}$
	5	$2.1117612255 \times 10^{-10}$	$1.1279989530 \times 10^{-10}$
	6	$1.8106651096 \times 10^{-11}$	$1.9737118563 \times 10^{-11}$
	7	$-1.6621201052 \times 10^{-12}$	$-1.1582995210 \times 10^{-12}$
	8	$-8.7167027819 \times 10^{-14}$	$8.6768075576 \times 10^{-14}$
	9	$1.0705140705 \times 10^{-14}$	$-2.7869365761 \times 10^{-14}$
	10	$-7.5223488683 \times 10^{-14}$	$3.3135927064 \times 10^{-14}$
	11	$-2.8387572286 \times 10^{-16}$	$2.0316367811 \times 10^{-16}$
	12	$-9.0639928622 \times 10^{-18}$	$-2.2565366439 \times 10^{-17}$
	13	$1.1380844737 \times 10^{-18}$	$-2.6006569868 \times 10^{-19}$

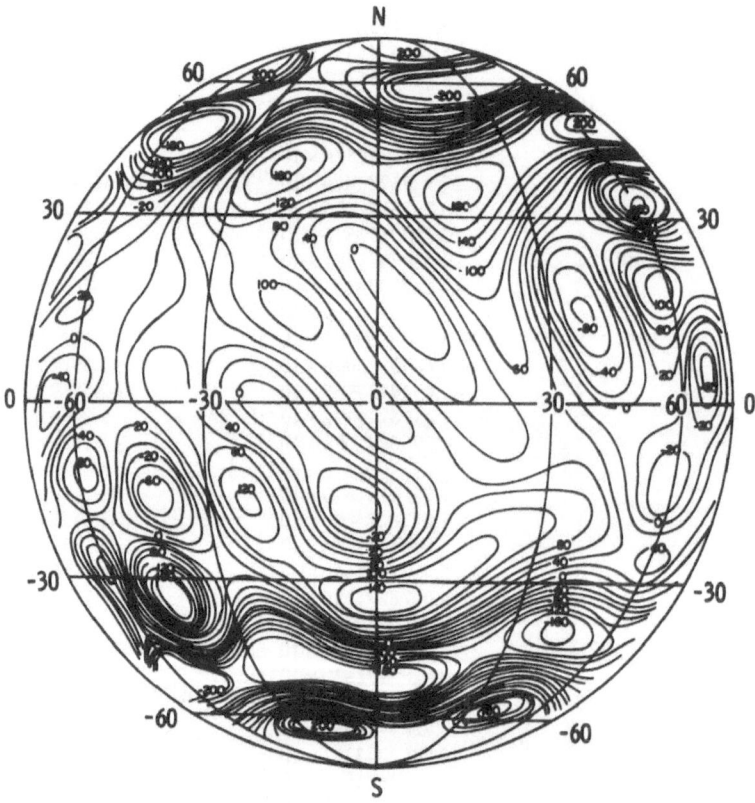

Fig. 4.3. A gravimetric map of the lunar near side, showing contours of the residuals of local gravitational acceleration (in milligals) from that expected for a spherical Moon, as deduced from the ensemble of the tracking of the motions of lunar orbiting satellites (after Michael and Blackshear, 1972). The solution for the residuals incorporated harmonic terms up to (and including) $n = 13$.

Fig. 4.4. The first detailed gravimetric map of the lunar surface, deduced from the observed perturbations of lunar orbiting satellites by Muller and Sjogren (*Science* **161**, 680, 1968). The map clearly shows the location of the principal 'mascons', and their relation to the circular maria.

and

$$C_{2,2} = \frac{B-A}{4m_{\mathbb{C}}r_{\mathbb{C}}^2}.$$ (4.10)

These two equations are obviously insufficient to specify the values of the three individual moments. However, on combining these with the values α, β, γ as defined by Equations (4.1) and numerically specified by Equations (4.4)–(4.6) from the study of physical librations, it is possible to specify the ratios $(A, B, C)/m_{\mathbb{C}}r_{\mathbb{C}}^2$ in absolute units. The latest such determination by Michael and Blackshear (1972) led to the value for the polar moment of inertia of

$$\frac{C}{m_{\mathfrak{c}} r_{\mathfrak{c}}^2} = 0.401 \pm 0.003 \, ; \tag{4.11}$$

with similar ratios for the other two moments.

The result reported since by Williams *et al.* (1973) or Bender *et al.* (1973) and based on the values of α, β, γ deduced from laser ranging of the Moon – namely,

$$\frac{C}{m_{\mathfrak{c}} r_{\mathfrak{c}}^2} = 0.395 \pm 0.005, \tag{4.12}$$

is not appreciably different from (4.11); and neither moment differs from the round value of $0.4 \, m_{\mathfrak{c}} r_{\mathfrak{c}}^2$ by an amount that can be regarded as significant. The physical meaning of this result for the internal structure of the Moon is clear, but its discussion in its wider context is being postponed for the subsequent chapter.

The recent work by Michael and Blackshear (1972) extended an analysis of the lunar gravitational field to harmonic terms of 13th degree. But even before their work, Muller and Sjogren (1968) analyzed the topographic distribution of orbital perturbations of the orbiting satellites for residuals which remained after the principal harmonics of low orders have been taken out; and in so doing made a remarkable discovery: namely, that of the existence of localized regions on the Moon exhibiting strong positive gravitational anomalies. Moreover, most of (though not all) these regions coincided with the locations of lunar circular maria – such as Mare Imbrium, Serenitatis, Crisium, Humorum or Nectaris.

A list of these regions has been compiled in the accompanying Table 4-3; and the respective gravimetric contours are shown on Figure 4.4. The anomalies themselves signify anomalous concentrations of mass (*'mascons'*, for short) of which there is no accumulation on the surface – in point of fact, the circular maria whose locations coincide with most of these 'mascons' are depressed below the level of the surround-

TABLE 4-3

Mascons associated with circular maria on the Moon

Name of the mare	Approximate location in latitude (β) and longitude (λ) of the centre of the mascon		Observed gravitational anomaly (in milligalls) after Muller and Sjogren (1969)
	β	λ	
Imbrium	$+38°$	$-18°$	220
Serenitatis	$+28°$	$+18°$	220
Crisium	$+16°$	$+58°$	130
Nectaris	$-16°$	$+34°$	120
Sinus Aestuum	$+10°$	$-8°$	80
Humorum	$-25°$	$-40°$	65
Humboldtianum	$+57°$	$+82°$	52
Orientale	$-20°$	$-95°$	52
Smythii	$-4°$	$+85°$	52
Sinus Iridum	$+45°$	$-31°$	-90

ing landscape (see again Figures 4.1 and 4.2). The observed rates of change in orbital motion during overflight by lunar spacecraft are, moreover, so abrupt as to indicate that the mascons responsible for such a behaviour must have fairly sharp rims and be relatively small in size (50–200 km across).

More detailed gravimetric work carried out with the aid of the S-band transponder by successive Apollo missions led, moreover, to a realization that smaller impact formations – in fact all ten of the 100 km craters (including Copernicus or Theophilus) that have been sampled so far – exhibit *negative* gravity anomalies, indicating local uncompensated mass deficiences. The same is true also of Sinus Iridum on the western shores of Mare Imbrium – the latter containing a positive mascon.

In terms of harmonic analysis, the mascons should emerge among surface harmonics corresponding to an index n greater than 20 – this is why the gravimetric map by Michael and Blackshear, terminating with $n = 13$ (cf. Figure 4.3) knew still little about them. Moreover – in order to produce the observed effects on the motions of overflying spacecraft – the mascons (positive or negative) can be located at only a shallow depth (50–100 km) below the surface; for deeper down the spacecraft could hardly feel them. Their masses appear to lie between 10–100 millionths of that of the whole Moon, and constitute overburdens of the order of 10^2–10^3 kg cm^{-2} on their substrates.

The existence of such sub-surface mascons – for which there is no obvious parallel on the Earth – imposes important constraints on the internal structure of the Moon and, in particular, on the rigidity of its outer mantle – a subject to which we shall now proceed to turn our attention.

References

Bender, P. L. *et al.* (12 co-authors): 1973, *Science* **182**, 229

Brown, E. W.: 1919, *Tables of the Motion of the Moon*, Yale Univ. Press, New Haven.

Eigen, J. M. and Hathaway, J. D.: 1967, in Z. Kopal and C. L. Goudas (eds.), *Measure of the Moon*, D. Reidel Publ. Co., Dordrecht, pp. 305–316.

Helmering, R. J.: 1973, *Moon* **8**, 450.

Kopal, Z.: 1972, *Moon* **4**, 28.

Kopal, Z. and Carder, R. W.: 1974, *Mapping of the Moon*, D. Reidel Publ. Co., Dordrecht.

Koziel, K.: 1967a, *Icarus* **7**, 1.

Koziel, K.: 1967b, in Z. Kopal and C. L. Goudas (eds.), *Measure of the Moon*, D. Reidel Publ. Co., Dordrecht, pp. 3–11.

Meyer, D. L. and Ruffin, B. W.: 1965, *Icarus* **4**, 513.

Michael, W. H. and Blackshear, W. T.: 1972, *Moon* **3**, 388.

Mills, G. A.: 1967, *Icarus* **6**, 131; **7**, 193.

Mills, G. A.: 1968, *Icarus* **8**, 90.

Mills, G. A. and Sudbury, P. V.: 1968, *Icarus* **9**, 538.

Muller, P. M. and Sjogren W. L.: 1969, 'Lunar Gravimetrics', COSPAR Meetings, Prague.

Rodionov, B. N. *et al.* (16 co-authors): 1971, *Kosm. Issled.* **9**, 450.

Schimerman, L. A., Cannell, W. D., and Meyer, D. L.: 1973, in *Trans. 15th General Assembly of IAU*, Sydney.

Sjogren, W. L.: 1967, in Z. Kopal and C. L. Goudas (eds.), *Measure of the Moon*, D. Reidel Publ. Co., Dordrecht, pp. 341–343.

Williams, J. G., Slade, M. A., Eckhardt, D. H., and Kaula, W. M.: 1973, *Moon* **8**, 469.

Wollenhaupt, W. R. and Sjogren, W. L.: 1972, *Moon* **4**, 337.

Wollenhaupt, W. R., Osburn, R. K., and Ransford, G. A.: 1972, *Moon* **5**, 149.

CHAPTER 5

INTERNAL STRUCTURE OF THE MOON

An investigation of the internal structure of any celestial body – be it star of a planet – possesses many features in common. Since it deals (by definition) with regions which are impenetrable to direct observation, it consists, in effect, of the construction of a model based on differential equations safeguarding the conservation of the mass, momentum and energy, for given type of boundary condition – such as the total mass, radius, rotational period and (for the stars) luminosity of the respective con-figuration – or other constraints furnished by the observed data. In point of fact, the stars and planets alike represent two classes of self-gravitating astronomical bodies differing mainly in mass (by several orders of magnitude); and all different external manifestations (such as the size, or luminosity of such bodies) stem directly from this source.

In what follows we plan to approach the main subject of our inquiry in successive stages of refinement, and draw first such simple and obvious conclusions as one can deduce from the facts already presented in the preceding chapters. The first concerns the mean density of our satellite. Its value of 3.34 g cm^{-3} deduced in Chapter 3 comes so close to the average density of lunar rocks collected on its surface as to leave little or no room for any appreciable increase inwards. In the case of our Earth, the mean density of rocks constituting the terrestrial crust (2.8 g cm^{-3}) turns out to be only a half of the mean density of the terrestrial globe (5.5 g cm^{-3}) – a comparison of which suggests strongly a considerable degree of compression.

For the Moon, nothing like this is suggested by the available data. In point of fact, the moments of inertia of the lunar globe correspond to what we should expect if the Moon were homogeneous throughout its interior. For, as is well known, the moment of inertia I about the centre of mass of a spherical configuration is related with the moments A, B, C of inertia about the principal axes by

$$I = \tfrac{1}{3}(A+B+C) = \tfrac{8}{3}\pi \int_0^{r_{\mathfrak{c}}} \varrho r^4 \, dr, \qquad (5.1)$$

which for the case of constant density ϱ yields

$$I = \tfrac{8}{15}\pi \varrho r_{\mathfrak{c}}^5 = \tfrac{2}{5}m_{\mathfrak{c}} r_{\mathfrak{c}}^2 \; ; \qquad (5.2)$$

and since, within the scheme of our approximation, $A = B = C$, the observed value of this quantity (cf. Equations 4.11 or 4.12) does not differ from that given by (5.2) within the limits of observational errors.

Such being the case, let us enquire next about the internal *pressure* to be expected inside a selfgravitating globe of lunar mass and size. If this mass were in hydrostatic equilibrium – an assumption which will have to be tested on its merits – the pressure $P(r)$ and the density ϱ at a distance r from the Moon's centre should, to a high degree of approximation* be related by the well-known equation

$$\frac{dP}{dr} = -g\varrho \tag{5.3}$$

of hydrostatic equilibrium, where the gravitational acceleration g inside a spherically-symmetrical configuration will be given by

$$g = G \frac{m(r)}{r^2}, \tag{5.4}$$

where G is the constant of gravitation; and $m(r)$, the mass interior to a volume of radius r, follows from the equation

$$\frac{dm}{dr} = 4\pi\varrho r^2 \tag{5.5}$$

as a consequence of the conservation of mass.

In order to solve these equations let us assume, to begin with, that the density ϱ is constant and equal to the mean density ϱ_m. If so,

$$m(r) = \tfrac{4}{3}\pi\varrho_m r^3, \qquad g(r) = \tfrac{4}{3}\pi G\varrho_m r; \tag{5.6}$$

and Equation (5.3) can be likewise immediately integrated to yield

$$P = \tfrac{2}{3}\pi G\varrho_m^2 (r_{\ell}^2 - r^2), \tag{5.7}$$

where the integration constant has been adjusted so as to make P vanish on the surface. The central pressure then comes out to be equal to

$$P_{\ell} = \tfrac{2}{3}\pi G\varrho_m^2 r_{\ell}^2 = 4.71 \times 10^{10} \text{ dyn cm}^{-2} \tag{5.8}$$

or 46 500 atm – a pressure easily matched in terrestrial laboratories, and exceeded in the Earth's mantle at a sub-surface depth of a mere 150 km.

How closely does the present state of the lunar globe actually come up to the requirements of hydrostatic equilibrium? The simplest and most obvious fact bearing on this matter is, of course, the very nearly spherical form of our satellite (cf. Chapter 4), demonstrating that the mass of our satellite has settled in the form of a configuration of minimum energy, of mean radius of 1738 km.

This observational result in complete agreement with consequences of hydrostatic theory; for even if the material constituting the lunar globe reacts to temporary impulses as an elastic solid, no common rock is known which could withstand indefinitely a pressure of the order of 10^4 atm prevailing in most part of the lunar

* Only minute effects of the small centrifugal force due to the Moon's slow axial rotation being ignored.

interior. Given a sufficiently long time (which is short in comparison with the age of our satellite) the material which on a short time-scale may behave as a solid is bound to get crushed under its own weight to settle to a form of minimum potential energy – which is a sphere. This is why not only gaseous stars like the Sun, but also 'solid' bodies like the major or terrestrial planets, are bound to become spherical in time, whatever their initial state may have been; a retention of non-spherical shape being the prerogative of those self-gravitating bodies (asteroids, meteorites) whose mass is too small to give rise to internal pressures capable of overcoming the crystal structure or molecular cohesion of solid state. If bodies so small were initially aspheric, they could remain so permanently and defy any efforts of self-attraction to establish hydrostatic equilibrium – as is, for instance, attested by observed light variations of the asteroids. The masses of even the largest of them – Ceres, Juno, Pallas, Vesta – are already too small for hydrostatic equilibrium to establish spherical shape; but the Moon, the first four Jovian satellites, or Mercury are well beyond this limit.

However, this agreement becomes noticeably worse when we come to compare the *momenta* of the lunar globe, as deduced from its physical librations, with their appropriate equilibrium values. In Chapter 4 we found that the most probable observed values of the ratios α, β and γ are as given by Equations (4.4)–(4.9); while their 'hydrostatic' values due to the combined rotational and tidal distortion should be given (cf., e.g., *Moon II*, p. 88) to

$$\alpha = 0.0000094,$$
$$\beta = 0.0000374, \tag{5.9}$$
$$\gamma = 0.0000280.$$

Could, perchance, such discrepancies be explained by the fact that the present dynamical characteristics of the Moon reflect a 'petrified' state of hydrostatic equilibrium at a time when our satellite was much nearer to us than it is at present? We may note that the observed value of β would be consistent with hydrostatic equilibrium if the Moon were $(17)^{1/3} = 2.57$ times closer to the Earth than it is now. However, at this smaller distance the Keplerian angular velocity ω_K of orbital revolution would have been $(17)^{1/2} = 4.12$ times larger than it is today – i.e., one month lasting only 6.63 of our days; and this would (in the case of continuing synchronism between rotation and revolution) correspond to a polar flattening of 0.28 km – while the tidal bulge directed to the Earth would have been almost 1 km in height. Neither this bulge, nor polar flattening, are borne out by the observed data – in particular, the front side appears to be compressed with respect to the mean moon-level (cf. Chapter 4).

Moreover, any such avenue of escape becomes completely ruled out when we come to consider the *ratio*

$$f = \frac{\alpha}{\beta} = \frac{B(C-B)}{A(C-A)} \tag{5.10}$$

of the respective differences of the momenta. For configurations in hydrostatic equilibrium, to the first order in superficial distortion we should have (cf., e.g., *Moon II*, p. 88)

$$f=\left\{1+\frac{36\,m^{\oplus}}{\omega^2 R^3}\right\}^{-1},$$

(5.11)

where $m_{\oplus}$ denotes (as in Chapter 3) the mass of the Earth; R, its distance; and ω, the Moon's angular velocity of axial rotation. If we identify (for synchronous rotation)* ω with the Keplerian angular velocity

$$\omega_K^2 = \frac{G(m_{\oplus}+m_{\text{(}})}{R^3}$$

(5.12)

of orbital revolution, Equation (5.11) will reduce to

$$f=\frac{m_{\oplus}+m_{\text{(}}}{4m_{\oplus}+m_{\text{(}}}$$

(5.13)

independently of the Moon's internal structure. As, moreover, the ratio $m_{\oplus}/m_{\text{(}}=81.302\pm0.001$ (cf. Chapter 3), the 'hydrostatic' value of the ratio f should be equal to 0.2523 or approximately one-quarter – in contrast with the value of 0.633 ± 0.011 deduced from the observations (Koziel, 1967a, b), or 0.642 ± 0.005 is deduced from laser rangings. This latter value is so much at variance with the requirements of hydrostatic equilibrium that the discrepancy must be considered real.

Therefore, in spite of the approximate prevalence of hydrostatic equilibrium in lunar interior as attested by the nearly spherical form of its globe, the motion of the Moon's globe around its centre of gravity disclosed the presence of small but unmistakable *departures* from this equilibrium; and the discrepancy between the observed and hydrostatic values of testifies to the extent to which the Moon must deviate from hydrostatic equilibrium somewhere in its interior. It is easy to see that this region cannot be sought very far from the surface; for regions deeper down become very ineffectual for this purpose. An outer shell of the Moon 200 km in depth contains (assuming uniform density) 30.7% of the Moon's mass, but accounts for 45.7% of its moments of inertia**. Moreover, at a depth of 200 km, the lithostatic pressure reaches (cf. Equation (5.7)) a value of about 10 kb – pressure at which rocks at moderate (let alone elevated) temperature can no longer remain rigid. Therefore, one would expect that matter in the lunar interior – up to (say) 1500 km from the centre – would be compelled by pressure to conform to the requirements of hydrostatic equilibrium; and that departures from this equilibrium could arise only in the outer zone where the lithostatic pressure has diminished well below 10 kb.

---

* Dynamical evidence is indeed conclusive (cf. Kopal, 1972; p. 33) that the Moon could not have departed from synchronism between rotation and revolution for any appreciable length of time in the past.

** The lunar mascons located within this zone are bound to contribute also to the effective moments of inertia of the lunar globe; but their effects are much too small to be of any real significance in this connection.

Further implications of these facts will be discussed in the concluding Chapter 10 of this book. In this place we wish to stress that additional information (albeit of more localized nature) on the Moon's departure from hydrostatic equilibrium has come in our hands with the discovery of the 'mascons', described already in Chapter 4 of the present volume. We mentioned that such mascons constitute localized lumps of denser material imbedded in the crust at a relatively shallow depth below the surface; and their existence alone provides us with a possibility of testing the rigidity of the lunar globe. As we shall explain in Chapter 8, we have every reason to believe that these mascons have been perched in their exposed positions for at least a few billion years (as the circular maria which closed in above them are of that age); and it was pointed out promptly and independently by Urey (1968) as well as Kopal (*Moon II*, pp. 203–204) that this could be the case only if the rigidity of the lunar interior exceeded that of the Earth's crust by 3 to 4 orders of magnitudes. On the Earth, similar formations exerting overburdens of the order of several hundred kg cm^{-2} on the underlying substrate would sink to depths at which their existence could no longer be detected within a few million years – that is why there are no mascons on the Earth (at least not with masses of the order of 50–$100 \times 10^{-6}\ m_\oplus$) – but greater rigidity (as well as lower gravity) has enabled them to survive near the Moon's surface for astronomically long intervals of time.

The continued presence of such mascons in the body of the Moon is bound to subject their surroundings to stresses which can be computed, and from which the rigidity of the respective layers can be estimated. This has been done, in recent years, most completely by Arkani Hamed (1972, 1973), who established that compression effects caused by the presence of the mascons can cause lateral density variations in the lunar crust amounting to 0.20 g cm^{-3}; and these do not get smoothed out till at sub-surface depths below 300 km. The stress-differences caused by the mascons are of the order of 100 bars within the first 400 km of the crust; and as the youngest mascons are not less than 3.3 billion years old (cf. Chapter 8), the lunar crust must be rigid enough to withstand such differences for at least such intervals of time.

The phenomena associated with the mascons permit us to reconnoitre, to some extent, the properties of the lunar mantle down to a depth of 300–400 km. A veritable royal road for probing the internal structure of the Moon as a whole has, however, been provided since 1969 by the combined output of *seismometers* installed on the Moon by the Apollo 11–17 missions, which have been provided us with an almost uninterrupted record of over 1000 individuals moonquakes up to this time.

The type of seismographs installed on the Moon by successive Apollo missions has already been described in Chapter 2, and the location of the first (Apollo 11) instrument *in situ* can be seen on Figure 2.6. This particular instrument functioned for only a few weeks; but those installed by Apollo 12, 14, 15 and 16 missions are still operative and will keep supplying us with new information on lunar seismicity for many years to come.

The characteristics of lunar seismograms differ very markedly from those typical for the Earth. The most striking aspect of lunar signals (see Figure 5.1) is their in-

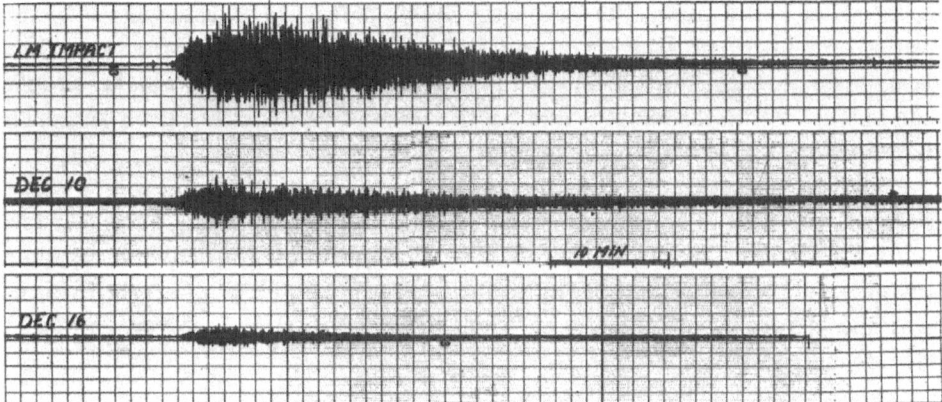

Fig. 5.1. Seismic signals from the Moon, received on the long-period vertical component seismometer from the impact of the Apollo 12 Lunar Excursion Module (LEM) on 20 November 1969, and from natural sources (presumably meteoritic impacts) in December 1969 (after Latham *et al.*, 1970).

ordinately long duration. Signals from the impacts of the S-IVB stages of the Apollo boosters, which on the Earth would be measured in minutes, on the Moon last for several hours. Lunar signals build up relatively slowly and increase gradually to a maximum, after which they decay even more slowly. The onset of direct shear waves is usually so indistinct that they can be but seldom inmistakably identified.

This, as well as other unusual characteristics of lunar seismic signals have been interpreted as resulting from intensive, but nearly loss-free, *scattering* of seismic waves in a heterogeneous layer which blankets the entire surface of the Moon (cf. Chapters 6 and 10); and the low seismic velocity of this layer makes the lunar surface act as an effective 'wave-guide'. Seismic waves generated, for example, by an object impinging on the Moon from the space are intensively scattered near the impact point; and only a part of their energy gradually leaks into the lunar interior. Since this interior transmits (at least through most part of it) both compressional and shear waves, a separation of the two becomes possible at far ranges.

Let us consider the propagation of seismic waves generated by a surface impact in greater detail. Seismic energy will radiate from the point of impact in the form of *body* (i.e., compressional and shear) waves which can travel through the lunar interior, and of *surface* (Rayleigh or Love) waves which travel along the surface. The waves of the latter kind will initially carry most part of the total energy, because of the 'wave-guide' effect of the regolith. It is only by a gradual conversion of surface to body waves that seismic energy of surface impacts gradually 'leaks' into the lunar interior.

Information on lunar structure down to a depth of about 100 km has been deduced primarily from seismic signals generated by impacts of the LM and S-IVB stages of successive Apollo missions. Information on lunar structure at greater depths can be obtained from an analysis of the seismic records of deep moonquake signals, as well as from large meteoroid impacts detectable at distances greater than 1000 km from the recording seismograph. The seismic signals of the two can be distinguished by the relative prominence of shear waves, and the differences in the signal rise times. The shear waves are much more prominent in the wavetrains of the moonquakes proper than in those due to the impacts, and the rise times of moonquake signals are much shorter than those accompanying impacts.

A number of locations emitting quakes on the Moon have been identified in the past four years, and their list has been compiled in the accompanying Table 5-1. The coordinates β and λ denote the latitude and longitude of the respective moonquake location; and the last column indicates the sub-surface depth at which the focus of the respective disturbances has been triangulated from seismic records secured at three or more locations.

TABLE 5-1

Selenographic locations of epicentres of recurrent moonquakes *

Latitude	Longitude	Depth of focus (in km)
23°3 S	28°3 W	850
15.0 N	25.0 E	700
44.8 S	53.4 W	800
9.2 N	8.4 W	610
30.2 N	2.9 E	800
22.3 N	30.2 W	940
22.8 N	61.1 E	860
7 S	139 E	800

* After Latham et al. (1973).

All moonquakes observed at these locations are, to be sure, tiny in terms of the terrestrial standards. With one possible exception, the largest of them are of the magnitudes between 1 and 2 on the terrestrial Richter scale; and the one best known (with epicentre at $\lambda = 28°3$ W, $\beta = 23°3$ S) entails recurrent energy release equivalent to the explosion of only about one kg of TNT per single event. It is only because the Moon is seismically so very quiet – dissipating only about 10^{15} erg of seismic energy per annum (equivalent to the explosive power of about 200 ton of TNT), in comparison with 5×10^{24} erg for the Earth – that moonquakes so small (corresponding to surface displacements of the order of 1 Å) and originating so deep in the interior can be recorded on the surface at all. The foci of most moonquakes so far identified are located at a depth between 610 and 940 km below the lunar surface – i.e., almost half way to the centre of the Moon – again in stark contrast with the Earth where almost all known quakes originate at a shallow sub-surface depth of no more than 2–3% of the terrestrial radius.

On the Earth, deep quakes are thought to be associated with lithospheric slabs drawn to greater depths by global convection in the mantle. On the Moon, where all evidence we possess precludes the existence of such movements, other explanations must be sought; and we shall probably not be far off the truth if we regard the 600–900 km zone inside the Moon with one in which maximum thermoelastic stresses are encountered.

Further implications of this suggestion will be followed up in a later part of this chapter. Before we do so, however, we wish to stress another outstanding feature of most moonquakes observed since 1969: namely, their recurrent nature. Moonquakes from each active focus have been established to recur in periods of 14 and 28 days – which are also the periods of the principal bodily tides raised on the Moon by the attraction of our Earth. Moreover, the nearly exact repetition of moonquakes signals from every focus over time intervals of many months requires that the focal region be small – 10 km in diameter or less – and immovable inside the Moon.

In addition to the monthly periodicity in the recurrence of moonquakes related with lunar tides, long-term variations in seismic activity at a given focus can also be correlated with known fluctuations of the tide-generating field of force (cf. Latham et al., 1971). Hence, tidal strain must contribute significantly to the total strain energy released in the form of moonquakes. In other words, tides may not only provide a trigger, but also the dominant energy source. Whether the strain itself is of thermal or gravitational origin remains yet a matter of speculation. It is, however, of interest to note that the positions of moonquake epicentres located so far (cf. Table 5-1) coincides quite closely with the rims of the major mascon basins.

In addition to lunar seismic activity described so far, Nakamura and his colleagues identified recently (cf. Nakamura, 1974) another type of moonquakes – events characterized by shallow epicentres (generally much less than 100 km below the surface), greater intensity (magnitude up to 4 on the Richter scale), and shorter duration (with seismic signatures much more akin to those of terrestrial earthquakes). Such events are rare – only 11 of these have been detected in the past four years – and may be due to tectonic shifts in the outer crust, or to impacts by unusually fast (interstellar?) meteors which penetrate through the regolith to bedrock. Whatever the case may be, a conspicuous lack of shallow moonquake activity (in contrast with the number of centres active at greater depth) precludes the accumulatic of strain in the Moon's outer parts which appear to be in steady-state condition.

What else are we entitled to conclude from the ensemble of the lunar seismic evidence which has, since 1969, come in our hands in such abundance? First, an analysis of seismic data from man-made impacts has established the existence of a lunar crust approximately 60 km thick, in which the velocity of seismic (compressional) waves is about 7 km s^{-1} – a velocity close to that expected for gabbroic anorthositic rocks (see Chapter 7). Beneath this crust, the velocity of compressional waves turns out to increase quite abruptly to 8.1 km s^{-1}, and remains approximately the same down to a depth of at least 1000 km.

By analogy with the Earth, this zone below the 'crust' can be referred to as the Moon's mantle; for the material within it continues to react to seismic disturbances as a solid (by transmitting both pressure and shear waves) of relatively high rigidity (corresponding to Poisson's ratio σ close to 0.25). However, in contrast with the

Earth – whose mantle amounts to 69% of the total terrestrial mass – the Moon's 'mantle' represents more than 90% of the total mass of our satellite; and as such renders the Moon largely a solid rock, which assumed spherical form because its rock strength could not permanently withstand the effects of gravitational self-compression.

The idea of present Moon as a solid spherical rock may, however, not represent a complete description of the central regions of our satellite. An analysis of the seismic data furnished by moonquakes with epicentres on the Moon's far side suggests (cf. Nakamura *et al.*, 1973) that, for seismic events located at a depth greater than 1000 km (i.e., about 700 km from the Moon's centre), shear waves seem to leave no longer any discernible signature in seismic messages emerging from such depth (see Figure 5.2). This signifies, in turn, that lunar material close to the centre ceases to behave as an elastic solid, and tends to become plastic (or even liquid). If so, this may also explain why the majority of recurrent moonquake foci are located at a depth not

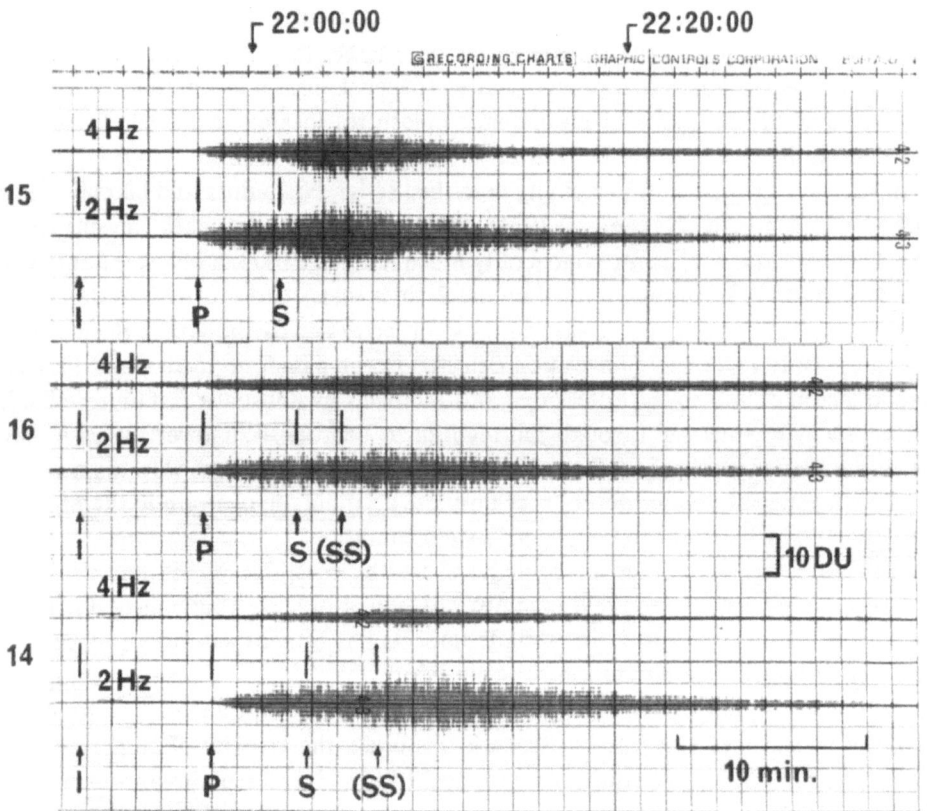

Fig. 5.2. Seismographic records of a meteoritic impact on the Moon's far side (at approximately $\beta = 30°$ N and $\lambda = 147°$ E, near Mare Moscoviense) on 17 July 1972 at 21 h 50 m s UT, secured from Apollo 14, 15 and 16 landing sites. Note that the characteristic S-wave bulge, visible at station 15, is missing at stations 14 and 16 – a feature which suggests at least a partial shielding of stations 14 and 16 by an interposed molten core which is opaque to the shear waves (after Nakamura *et al.*, 1973).

very far above the level at which a transition in state of the lunar material may become effective. But further information will be necessary before such a suggestion can be placed on a more solid footing.

Be it as it may, however, the available seismic evidence leaves no room for doubt that the surface of the Moon is covered with a highly heterogeneous layer through which seismic waves propagate with little damping, but a great deal of scattering. It is the presence of this layer which accounts for most part of the marked differences between the lunar and terrestrial seismic signals; and its structure is no doubt a consequence of the cratering processes – mainly by external impacts – which the Moon has experienced since the time of its formation.

What is the absolute *temperature* of the lunar interior in the light of the Apollo results? While a fuller discussion of the thermal history of our satellite is being postponed till the concluding Chapter 10 of this book, let it be mentioned already now that – as most other bodies of the solar system – the Moon very probably originated by an agglomeration of solid particles at relatively low temperatures (say, a few hundred degrees Kelvin). Once, however, this process has been accomplished (in a time of not more than a few million years), additional heat must be continuously generated throughout the Moon's interior by a decay of such traces of natural radioactive elements (e.g., potassium K40, thorium Th232, or the two isotopes of uranium U235 and 238) as were present in the primordial lunar material. Therefore – as in the stars – the thermal energy produced now inside the Moon is also due to nuclear transformations. But whereas, inside the stars, the reactions concerned are of the fusion type, and the prevalent conditions are sufficiently extreme for the rate of these reactions to be influenced by the local density and temperature, all exothermic reactions occurring now in the Moon are restricted to spontaneous disintegration which proceeds at a constant rate supremely oblivious of the ambient conditions (for the relevant data, cf. the accompanying Table 5-2).

The internal temperature prevalent inside the Moon at any time as a result of the output of such reactions should then reflect a balance between the amount of the production of radiogenic heat in the interior, and its loss through the surface by radiation into the surrounding space. The amount of heat generated in the interior should be proportional to the volume of the Moon; and its loss, to the area of its surface. For spherical configurations such as the Earth or the Moon, a ratio of the

TABLE 5-2

Principal cycles for the production of radiogenic heat

Disintegration chain	Half-life (in 10^9 yr)	Total heat production (in 10^{20} erg per kg of mother substance)
U 238 – Pb 206	4.51	1.922
U 235 – Pb 207	0.713	1.856
Th 232 – Pb 208	13.9	1.655
K 40 – Ar 40	1.306	0.171

volume to the surface should be proportional to the radius of the respective cosmic body. Therefore, a planetary body of the size of the Earth should be able to bottle up in its interior four times as much heat as the Moon – an expectation which seems indeed to be borne out by the evidence we are going to discuss.

The temperature prevalent on the surface of our satellite is essentially controlled by insolation – this is why it exhibits diurnal variations ranging between 80–400 K, described more fully in Chapter 20 of *Moon II*. The heat flux from the lunar interior – at depths beyond the reach of the diurnal heat wave – was measured only twice by the Apollo missions – in the proximity of the Hadley rille by Apollo 15 in July 1971, and near the Taurus-Littrow landing place of Apollo 17 in December 1972. Near the Apollo 15 landing place, the heat flux in the outermost $1\frac{1}{2}$ m of the lunar crust proved to be equal to approximately 33 erg cm^{-2} sec (cf. Langseth *et al.*, 1972); and 27 erg cm^{-2} s^{-1} in the Taurus-Littrow region (Langseth *et al.*, 1973). Such flux should be sufficient by itself to raise the surface temperature from 0 to 22–27 K; but in comparison with the flux due to insolation its effects on the lunar infrared emission should be observationally insignificant.

These results are interesting because they represent the first direct measurements of the heat flux from the lunar interior anywhere on the Moon; but their interpretation requires caution. On the Earth, the heat flux measured in different parts of the world may differ by 30–40% from the mean; and if the source of this heat flux is radioactive, conspicuous variations in lunar surface radioactivity measured by Apollo 15–16 from orbital altitudes (cf. Chapter 7) should lead us to expect on the Moon much greater local variations.

And an even greater caution should be exercised in considering a significance of the 'selenothermic degree' of 1.3–1.7 deg m^{-1} as measured by Apollo 15 and 17 astronauts by devices reaching no more than 1–2 m below the surface. Such a degree may merely reflect local thermal vagaries of the particular spot, and is obviously not extrapolable to much greater depths; for it it were, the temperatures at which rocks melt would be attained at levels where – as we already learned from seismic evidence – the rocks are certainly solid. Indeed, it is almost the same situation as we encountered previously on the Earth. Here too the 'geothermic degree' near the surface – amounting to some 20–30 deg km^{-1} – is so steep that if it were linearly extrapolable inwards, temperatures at which rocks melt would be reached at subsurface depths no greater than 50 km; and internal temperatures far in excess of the melting point of the constituent material would be reached not only in the Earth's liquid core, but also throughout most part of the mantle.

The reason why the 'geothermic degree' is so steep is very probably the anomalously high concentration of heat-producing radioactive elements (listed in Table 5-2) in the Earth's crust. When the first lunar rocks were returned to the Earth by successive Apollo missions, it transpired that the crust of the Moon is similarly enriched with radioactive material – to a degree which cannot persist much deeper down if the Moon's mantle is to behave as a solid down to a depth of at least 1000 km (as indicated by seismic evidence discussed earlier in this chapter). In the case of the

Earth, geochemists are considering the possibility that the requisite chemical differentiation may have been brought about in the course of a slow circulation of the material through creeping motions in the terrestrial mantle. Since, however, the lunar crust is much more rigid than the terrestrial mantle, chemical differention caused by mass motions cannot be readily applied to the Moon. On the other hand, a possibility cannot be ruled out that this phenomenon – both on the Earth as well as on the Moon – can represent the effects of a last stage of nucleogenesis in the solar system, when the youthful Sun may have sprinkled the surfaces of new-born planets with a flux of neutrons emitted before the large-scale convection of its Hayashi stage died out completely in its interior.

Astronomers are still far from unanimous about the physical properties of the last stage of solar collapse to the Main Sequence. Deep convective zone of the Sun (which seems characteristic of stellar models at that stage) could have, however, engulfed the pristine surfaces of new-born planets in a 'solar wind' some 10^7 times more intense than at the present time; and this wind could have contained an appreciable component of neutrons. An absorption of these neutrons by the surface material of planets in the last stage of their formation may indeed have produced a short spell of nucleogenesis, of which the anomalously high abundance of uranium, thorium and certain other heavy nuclides may represent a surviving vestige.

If this was the case, then we should expect the surface of Mercury (and, to a lesser extent, Venus) to be more radioactive than the crust of the Earth and the Moon; and Mars to be less radioactive. Positive evidence that this is the case is so far lacking; but such indications as we possess are indeed consistent with such a hypothesis. Thus the recent mission of Mariner 10 to Mercury detected in its exosphere a higher concentration of helium than could be accounted for by solar influences; and surface radioactivity could be one explanation. Furthermore, the Venus landers of recent years detected a γ-ray emission from its surface indicative of super-terrestrial radioactivity. On the other hand, the testimony of Mars seems so far less conclusive: its γ-ray emission is likewise appreciable, but may be due in part at least to an interaction of its atmospheric constituents with the solar cosmic rays. Space missions of the years to come will no doubt confirm or disprove our hypothesis. But until they have delivered their verdict, the possibility that the youthful Sun may have sprinkled the surfaces of new-born planets with a 'neutron gun' before large-scale convection died out in its interior should be kept in mind.

But be it as it may, our present conclusions about the thermal state of the deep interior of the Moon must still rest essentially on indirect evidence. The seismic profile of the lunar globe – outlined in earlier parts of this chapter and disclosing the ability of the lunar material to transmit shear waves down to a depth of at least 1000 km – implies that, down to this depth, the prevalent temperature must be below the melting point of the constituent rocks (which is known to be close to 1200 °C). Indications are that, below this level, the lunar material becomes plastic – or even possibly molten – which, if true, would necessitate the central temperature of the Moon to be at least 1300°–1400 °C; but still very much less than is the case for the

Earth. This was, perhaps, only to be expected; because the much larger terrestrial mass can bottle up in its interior a proportionally greater amount of radiogenic heat; so that – for once – theory and observations seem to be at least qualitatively in agreement.

Another constraint on the temperature of the Moon's interior is provided by the observed magnetic interaction of the lunar globe with the solar wind – or, rather, a lack of it. A systematic study of this subject commenced in 1967, with the launch of Explorer 35 lunar satellite which we described already in Chapter 1. Magnetometric observations secured aboard this spacecraft have demonstrated that – unlike the Earth – the Moon does not produce any shock wave in interplanetary medium through which it is moving with supersonic velocity of orbital motion. The Moon does not accrete interplanetary magnetic field lines carried by the solar wind; but, very largely, casts a geometrical shadow in the plasma in anti-solar direction. Moreover, a diffusion of interplanetary magnetic field lines through the lunar globe discloses that the Moon behaves, to this extent, like an insulator rather than a semi-conducting body.

The upper limit to the global electrical conductivity, consistent with the observed lack of interaction, has been found to be of the order of 10^{-5} mhos m^{-1}. We shall, moreover, see in Chapter 7 that the material so conducting are essentially silicate rocks, whose conductivity is a well-known function of absolute temperature. If we are to reconcile the physical properties of such rocks with the orbital magnetometric evidence, it can be shown (cf., e.g., Sonett et al., 1971a, b, 1972, 1973) that the lunar internal temperature down to a depth of at least 700–800 km should be below 1000 °C. Such temperatures are sufficiently low for the lunar mantle to possess a degree of rigidity consistent with the seismic data; and thus, to this extent, different types of observations converge to the same picture.

The work of Explorer 35 provided us also with another piece of evidence on the physical properties of the lunar globe, concerning the *magnetic field* of our satellite – or rather, again, the lack of it. That any magnetic field the Moon may possess must be very weak came out already of the experiments performed by Luna 2 in September 1959 (cf. Dolginov et al., 1960, 1962) which indicated that the strength of the lunar dipole field – if any – did not exceed some 30 γ's (i.e., 3×10^{-4} G); and although some questions had then arisen as to the possible effects of solar wind on this measurement (cf. Neugebauer, 1960), the essential result was fully confirmed by subsequent work of Luna 10 (Dolginov et al., 1967) as well as Explorer 35 (Sonett and Colburn, 1967). As a result of all this work we know now that the general dipole field of the Moon – if any – does not exceed a few γ's in strength; and that the total magnetic moment of the lunar globe is probably less than 10^{-6} that of the Earth.

With the arrival, on the Moon, of the surface and subsatellite magnetometers of successive Apollo missions a whole spectrum of new facts bearing on lunar magnetism came to light. Thus sub-satellites of Apollo 15–17 missions, orbiting at altitudes substantially lower than that of Explorer 35, disclosed the existence of local magnetic fluctuations amounting to 20–30 γ's. However, already the Apollo 15 orbital magnetometer disclosed (cf. Coleman *et al.*, 1972; Sharp *et al.*, 1973) that such fluctuations are strongly correlated with specific formations on the lunar surface overflown by the magnetometer. The message of this fact is clear: namely, that the source of such fields must be only skin-deep as far as the lunar globe is concerned, and cannot have anything to do with its deep interior.

The samples returned from the lunar surface fully confirmed this; for it transpired that the carriers of this field are the brecciated rocks (constituting some 90% of the overlay of the lunar continents), which acquired their present characteristics as a result of multiple impacts. It is, therefore, more than probable that these breccias were magnetized by impacts – i.e., essentially an external influence – and that their present magnetic properties have nothing to do with those of the lunar interior – now or at any time in the past.

The same is *not*, however, true of the magnetic properties of lunar crystalline rocks; and of their remanent magnetism. This magnetism is stable, and suggests the prevalence of fields of 10^2–10^3 γ's throughout the periods during which the crystalline rocks solidified. Such fields would be 50 to 500 times less intense than the present magnetic field of the Earth, but 20 to 200 times more so than the fields carried by the present-day solar wind. Whether or not the mechanism which generated such fields in the past was internal (dynamo in a liquid metallic core, which had since become extinct) or external (such as greater proximity of the Earth, or anomalous solar wind), cannot be decided at the present time. Both these alternatives appear to be unlikely on many grounds; but nothing more plausible can so far be advanced in their place. The origin of the fields which gave rise to the observed remanent magnetism of lunar crystalline rocks continues to represent another unsolved problem of lunar studies.

References

Arkani Hamed, J.: 1972, *Moon* **6**, 100, 112, 135.

Arkani Hamed, J.: 1973, *Moon* **7**, 84.

Coleman, P. J., Schubert, G., Russell, C. T., and Sharp, L. R.: 1972, *Moon* **4**, 419.

Dolginov, Sh. S., Yeroshenko, E. G., Zhuzgov, L. I., Pushkov, N. V., and Tyurmina, L. O.: 1960, *Iskustv. Sputniki Zemli* **5**, 149.

Dolginov, Sh. S., Yeroshenko, E. G., Zhuzgov, L. I., Pushkov, N. V.: 1962, in Z. Kopal and Z. K. Mikhailov (eds.), 'The Moon', *IAU Symp* **14**, Academic Press, New York and London, pp. 45–62.

Dolginov, Sh. S., Yeroshenko, E. G., Zhuzgov, L. V., and Zhulin, I. A.: 1967, *Geomag. Aeronom.* **7**, 436.

Kopal, Z.: 1972, *Astrophys. Space Sci.* **16**, 3.

Koziel, K.: 1967a, *Icarus* **7**, 1.

Koziel, K.: 1967b, in Z. Kopal and C. L. Goudas (eds.), *Measure of the Moon*, D. Reidel Publ. Co., Dordrecht, pp. 3–11.

Langseth, M. G., Clark, S. P., Chute, J. L., Keihm, S. J., and Wechsler, A. E.: 1972, *Moon* **4**, 390.

Langseth, M. G., Keihm, S. J., and Chute, J. L.: 1973, in *Apollo 17 Preliminary Science Rept.* (NASA SP-330), 9–1.

Latham, G., Ewing, M., Press, F., Sutton, G. *et al.*: 1970a, *Science* **167**, 455.

Latham, G., Ewing, M., Press, F., Sutton,G. *et al.*: 1970b, *Science* **170**, 620.

Latham, G., Ewing, M., Press, F., Sutton, G. *et al.*: 1971, *Science* **174**, 687.

Latham, G., Ewing, M., Dorman, J., Lammlein, D., Press, F., Toksöz, N., Sutton, G., Dunnebier, F., and Nakamura, Y.: 1972, *Moon* **4**, 373.

Latham, G., Ewing, M., Dorman, J., Nakamura, Y., Press, F., Toksöz, N., Sutton, G., Dunnebier, F., and Lammlein, D.: 1973, *Moon* **7**, 396.

Nakamura, Y.: 1974, in *Lunar Science V*, Pergamon Press, in press.
Nakamura, Y., Lammlein, D., Latham, G., Ewing, M., Dorman, J., Press, F., and Toksöz, N.: 1973, *Science* **181**, 49.
Neugebauer, M.: 1960, *Phys. Rev. Letters* **4**, 6.
Sharp, L. R., Coleman, P. J., Lichtenstein, B. R., Russell, C. T., and Schubert, G.: 1973, *Moon* **7**, 322.
Sonett, C. P. and Colburn, D. S.: 1967, *Nature* **216**, 340.
Sonett, C. P. and Mihalov, J. D.: 1972, *J. Geophys. Res.* **77**, 588.
Sonett, C. P. and Runcorn, S. K.: 1973, *Moon* **8**, 308.
Sonett, C. P., Colburn, D. S., Dyal, P., Parkin, C. W., Smith, B. F., Schubert, G., and Schwartz, K.: 1971a, *Nature* **230**, 359.
Sonett, C. P., Mihalov, J. D., Binsack, J. H., and Moutsoulas, M. D.: 1971b, *Science* **171**, 892.
Urey, H. C.: 1968, *Science* **162**, 1408.

CHAPTER 6

MORPHOLOGY OF LUNAR FORMATIONS

In the preceding chapter of this book we paid a brief visit to the interior of the Moon, in order to get acquainted with the principal features of its structure. In the present chapter we shall emerge from the uninviting internal depths to the surface of our satellite, in order to discuss some problems arising in connection with the observed stony sculpture of this surface, and its origin.

A mere glance at an almost bewildering array of such formations of all sizes – as shown on many photographs accompanying this book – makes it appear unlikely that all of them originated in the same way or at the same time; and a more detailed analysis of their features suggests that this suspicion is probably well founded. In fact, the most reasonable approach to this problem can be made if we ask ourselves: What are all the processes which could have conceivably cooperated in shaping up the surface of our satellite? And once we thus formulate our problem, we find ourselves facing two principal contending theories of crater origin: namely, the external theory – invoking the effects produced by *impacts* of other celestial bodies (asteroids, meteorites, or comets) on the lunar surface – and the alternative theory relying on the *internal processes* connected with convection, gradual defluidization and degassing of the lunar globe consequent upon its build-up of internal heat due to spontaneous decay of radioactivity elements, or any other activity which could be loosely termed as 'volcanic'. In point of fact, *the entire surface of the Moon must be regarded as the outer 'boundary condition' of all internal processes which may have been going on in the lunar interior since the origin of our satellite, as well as an 'impact counter' of external events which may have visited it from outside.* In no other sense can an interpretation of the lunar surface possess any physical meaning.

A general survey of the type of formations which can be produced on the Moon by either action has already been given in Chapters 16 and 17 of the *Moon II*, the contents of which need not be repeated in this place. Instead, we propose to take for granted that our reader is at least to some extent familiar with the principal types of lunar surface formations; and our main aim will be to bring such a reader up to date in the contributions made to our acquaintance with the lunar surface made by spacecraft of the past four years.

Our reader is doubtless familiar in outline with the principal types of the lunar ground – the continents and the maria – as well as with the principal types of formations which disfigure the ground: namely, the so-called 'craters'. The origin of these formations – regarded as typically lunar until they were discovered in 1964 to pockmark similarly the surface of Mars and, in 1974, that of Mercury – has been widely discussed for a better part of two centuries; the pendulum of opinion swinging back and forth between 'volcanists' who saw in them homologues (on larger scale) of terrestrial volcanic structures, and scientists who saw in cratering by external impacts the principal clue to a proper understanding of the enigmatic hieroglyphs of the lunar face.

As long as this face could be observed only at a distance by means of the telescopes, room existed indeed for alternative interpretations, depending sometimes on the imagination of the protagonists. It started already in the earliest days of telescopic observations, which led Galileo Galilei to recognize the face of the Moon to be 'full of inequalities' – a view Galileo had to defend in the face of strenuous opposition by the Aristotelians, according to whose lights the Moon (as a celestial body) had to possess a perfectly smooth surface. Father Kley (Clavius), one of the leading astronomers of that day, attempted to reconcile the telescopic evidence with the Aristotelian pre-conceptions by assuming the surface roughness to be overlaid by a smooth but perfectly transparent crystalline substance, the surface of which met the Aristotelian requirements. "Really this is a beautiful flight of the imagination" commented Galileo (1611); "The only thing lacking in it is that it is neither demonstrated, nor demonstrable." We quote this because a similar comment could be applied to only too many other views on the Moon expressed in more than three centuries, from Galileo to our own times.

Fifty-seven years after Galileo's observations of the Moon, Robert Hooke (1635–1703) – who, like his younger contemporary Isaac Newton (1642–1729), was also interested in the Moon – dropped bullets into a pipe clay and water mixture and, behold, saw formations arise which one could call 'impact craters'. But being an inquisitive soul, Hooke did not stop there; but as he tells us in his *Micrographia* (1667) he also boiled a mixture of powdered alabaster with water, and observed that this too produced transient crater-like structures on the surface of the liquid. Thus he started a very interesting controversy about the origin of lunar craters, which has not been completely settled ever since to everyone's satisfaction.

Hooke himself rejected the impact analogy because, "it would be difficult to imagine whence those bodies should come." This opinion was formed, to be sure, more than a century before the extra-terrestrial origin of any meteorite was recognized; and by that time a volcanic origin of lunar craters appears to have been almost universally accepted by the leading scientists of the time. Thus a volcanic origin had been championed by Immanuel Kant (1785); and William Herschel even reported in 1787 to have seen what he believed to be observations of volcanic eruptions on the Moon. The impact hypothesis for the origin of the lunar craters was revived, to be sure, by Gruithuisen (1829), who also appears to have anticipated the planetesimal accretion hypothesis for the origin of the Moon. Nevertheless, many of Gruithuisen's views (such as his speculations about the inhabitability of our satellite) were rather extreme – a fact which did not increase the credibility of his more reasonable views – with the result that most of the principal selenologists continued to accept some form of volcanism as the crater-forming process on the Moon without much reservation.

The pendulum of the scientific thought on the origin of lunar surface features did not begin to swing away from volcanism and back to impacts until the work of the American geologist G. K. Gilbert (1893), who reviewed the characteristics of lunar craters together with those of the various types of terrestrial volcanoes; and concluded that the differences in form (and, to a lesser extent, in size) between the respective lunar and terrestrial objects were so great that a volcanic origin for the lunar craters seemed improbable. From this Gilbert went on to develop a consistent impact hypothesis for the origin of the lunar craters, which was based on some acute telescopic observations of the Moon as well as upon laboratory experiments.

Gilbert's views appeared, however, to be too advanced for his time; and by the end of the 19th century the consensus among a large majority of astronomers and geologists was still firmly in favor of volcanic origin of lunar craters. It was not till throughout the first half of the 20th century that the opinion began to shift significantly towards external impacts – mainly among the geologists – a fact which led, in 1926, the American geologist W. M. Davis to record (in his obituary of Gilbert) candidly that

...It has been remarked that the majority of astronomers explain the craters on the Moon by volcanic eruption – that is, by an essentially geological process – while a considerable number of geologists are inclined to explain them by the impact of bodies falling upon the Moon – that is, by an essentially astronomical process. This suggests that each group of scientists finds the craters so difficult to explain by processes with which they are professionally familiar that they prefer to take recourse to a process belonging to another field than their own, with which they are probably imperfectly acquainted and with which they therefore feel freer to take liberties. (Davis, 1926).

Since the time when these words were written, the principal protagonists of this great debate have almost completely exchanged position. With few significant exceptions, the majority of astronomers (who, in the meantime, became better acquainted with the particulate contents of interplanetary space, came to regard external impacts as the principal cratering mechanism operative on the lunar surface – a view exemplified, e.g., by Baldwin's book on the *Face of the Moon* (Baldwin, 1949) – while the geologists gradually reverted to champion purely internal processes – mainly volcanic – to account for the same outcome – a view carried to an extreme, e.g., in Spurr's *Geology Applied to Selenology* in four volumes (Spurr, 1944–1949); and it was this latter view which seems to have attracted greater following among more casual onlookers.

It has been a matter of some curiosity to the author of this book why should the 'volcanic' hypothesis of the origin of the lunar craters have enjoyed in recent years so much more support among the geologists (and an emotional support among laymen) than the 'impact' hypothesis favoured, in general, by the astronomers and physicists. The reason of this imbalance probably goes back to a different degree of familiarity with the homologues of the respective formations on the Earth. It is true that volcanoes are not – thanks God – particularly abundant per unit area of the terrestrial surface; but in active state they cannot help attracting attention in any part of the world; and, moreover, their recurrent nature may help them to remain conspicuous landmarks for millions of years (while the lava flows disgorged during their outbursts may be recognizable after 10–100 times as long).

On the other hand, impact phenomena are virtually instantaneous (i.e., occur on a time-scale of 10^{-7} of a year, instead of 10^7 yr as for the volcanoes); and their very abruptness makes them so devastating that no nearby witness could survive the experience at a close distance. The eruption of Krakatoa in 1883 – gruesome as it was – should still be a child's play in comparison with a (say) asteroidal impact that would release the same amount of energy within seconds rather than days. Moreover, once an impact crater has been formed, it becomes vulnerable to erosion which will obliterate it with time intervals of the order of 10^5–10^6 yr – with (unlike volcanoes) no means of regeneration. Many craters due to impacts have now been identified in different parts of the world, but virtually all are less than a million years old; while volcanoes 20–30 million years old remain quite conspicuous, and many are still active. Therefore, to appreciate the effects of past impacts requires a greater power of abstraction than to visualize a volcanic explosion – an exercise to which the empirically-minded geologists are less inclined than the astronomers who have had to practice it much more intensively out of sheer necessity.

The comments are not recorded out of any professional malice, but to underline some regret at the outcome of the influence which geologists so motivated exerted on the selection of the landing sites of the Apollo missions. In particular, in the latter part of the programme (Apollo 15–17), emphasis has increasingly been placed on a search for 'young' volcanic features on the Moon which the first three missions conspicuously failed to detect – a quest in which the occurrence of 'dark spots' in certain parts of the lunar surface has played the role of a mirage. Needless to say, also this mirage proved to be deceptive, and no signs of recent volcanism have been detected anywhere on the Moon so far. As we shall detail more fully in Chapter 8, all possible lunar 'lava flows' are more than 3000 million years old; yet it was mainly this quest which sent Apollo 17 to land at a locality for which its orbital track of approach had largely to duplicate that of the preceding mission 15.

To an astronomer, the great age of the lunar surface renders, of course, our satellite an object of all the greater interest – because its features constitute the link with a more distant past (of which otherwise we should have no direct source of information). Besides, no astronomer would expect two celestial bodies like the Earth and the Moon – with masses differing by two orders of magnitude – to have much in common in their internal structure or surface characteristics. Many geologists seem, however, to think otherwise; and will not, we suspect, feel really like at home on the Moon unless they can identify there the whole range of terrestrial homologues which they are accustomed to see on the Earth.

On the other hand, to the astronomers it has become, by the middle of the 20th century, increasingly evident that while volcanic activity *may* have occurred on the Moon on some scale in the past, cratering by external impacts *must* have been operative. For the Moon does not orbit in empty space, and is not protected by any bumper from impacts of bodies whose orbits happen to intersect its own. And, we may add, such bodies were no doubt very much more numerous in space in the early days of the history of the solar system than they are today.

The advent of lunar spacecraft of the 1960's – in particular, of the photographic Lunar Orbiters (cf. Chapter 1) – extended our acquaintance with the detailed structure of the lunar surface immeasurably. However, it was not till 1968 that the 'first-generation' negatives of the photographs of the lunar surface taken from a close proximity were actually returned to the Earth; and with them a vast amount of high-resolution data on the ground overflown by successive Apollo missions. The highlights of this effort have been undoubtedly the Apollo 15–17 missions, whose Command Modules were equipped with 'panoramic' as well as 'metric' cameras, to photograph a belt of ground more than 100 km in width, with a resolution of 1–10 m from an orbital altitude close to 100 km. The photographic star performer was without doubt Apollo 15 mission, which brought back several thousand photographs of excellent quality. The photographic output of Apollo 16 was degraded somewhat by difficulties with focussing of the optics, while Apollo 17's track overlapped partly with that of Apollo 15. In what follows, we shall reproduce a number of photographs obtained by these missions to demonstrate features exhibited by them which throw new, or additional, light on our previous knowledge of the subject as reflected in the previous edition of our book (cf. Chapters 16–17 of *Moon II*).

As far as the lunar craters are concerned, we now know that the principal cause of their presence on the lunar surface in such prodigious numbers are external impacts. Already prior to 1969 the case for such a thesis was strong; and since then it has become overwhelming. All craters visited by men on the Moon between 1969–1972 proved to be impact formations. Moreover, the prevalence of impact cratering also everywhere else is attested by the abundance of brecciated rocks, showing distinct effects of shock metamorphism. The impact origin of such large craters as Copernicus (see Figures 1.27 or 2.16), long conjectured, has now been attested by a determination of the age of its formation by radioactive methods (cf. Chapter 8). An application of these to rocks from the Copernican ejecta, collected by Apollo 14, shows that this crater was formed some $2\frac{1}{2}$ billion years after the rocks of the surrounding plains of the Oceanus Procellarum have solidified (cf. Chapter 8). The same is more than probable for such craters as Theophilus (Figures 8.1 and 8.5) or Tycho (Figures 1.30–1.32), for morphological reasons discussed already in Chapter 16 of the *Moon II*.

The Apollo photography of the lunar surface augmented our previous knowledge in several important respects; and the descent of men on the surface itself extended this acquaintance down to features of, not millimetres, but microns in size. Prior to the end of 1968, all views of the lunar landscape recorded from the vantage points of cis-lunar spacecraft had to be televised for re-assembly on the Earth – a process in which a certain amount of valuable information got irretrievably lost. The Russian Zond 6 and the American Apollo 8 of November–December 1968 transported to the Earth the first original photographs of the lunar landscape which no longer had to be televised (for their examples, see Figures 2.13–2.14 or 2.18) and systematic

Fig. 6.1. An oblique photograph of the crater Humboldt, taken with the metric camera of Apollo 15 mission from an altitude of 115 km above the lunâr surface (NASA official photograph).

orbital photography by the panoramic as well as metric cameras from the command modules of Apollo 15–17 missions marks a milestone in our acquaintance with wide belts of the surface of our satellite overflown by the respective missions (for their examples see, e.g., photographs of the craters Humboldt or Aristarchus-Herodotus, reproduced on the accompanying Figures 6.1 and 6.2).

What did we learn from this new evidence about the significance of different types of lunar formations? The subject of such an inquiry belongs to the domain of photogeology; and the amount of work currently in progress on the basis of all new evidence is truly enormous; so that only some highlights can be briefly mentioned in this place.

First, we learned to appreciate more fully the role of *subsidence*, which mechanical disturbances (mainly external) may bring about in the loosely-packed lunar crust.

Fig. 6.2. A photograph of the Aristarchus-Herodotus craters with Schröter's canyon and adjacent plains of Oceanus Procellarum to the south (above), taken with the metric camera of Apollo 15 mission from an altitude of 107 km above the lunar surface (NASA official photograph).

In fact, most craters less than 1 km in size (and many larger ones) are essentially depressions in the ground, with only a suggestion (if any) of upturned rims (cf. a photograph of the crater Römer reproduced on Figure 6.3). Moreover, the loose nature of the ground in which such formations were 'scooped' by impacts is attested by a great number of landslides, manifest on many such photographs as that of the crater Bessel shown on the accompanying Figure 6.4.

In extreme cases, evidence of subsidence is seen even at the very centre of such crater formations. As is well known, many large craters of impact origin – such as Copernicus (Figure 2.16), Theophilus (Figure 8.1) or Tycho (Figure 1.30) are characterized by the presence of a 'central mountain' – usually a group of hills approaching (though seldom attaining) the altitude of crater rims. On a photograph of the crater Timocharis shown on Figure 6.5 we see an opposite phenomenon: namely, an open cavity in the position where the central mountain should be – as though this mountain disappeared into the ground through the funnel ('dimple crater') left behind it.

That subsidence processes have played a greater role in shaping up the fine structure of the lunar surface than it appeared before 1969 has also been attested by new information bearing on the domes, rilles and wrinkle ridges. Both these characteristic surface features are obviously of internal origin, as by-products of the processes which accompanied the formation of the mare-basins (see Figures 6.6–6.9) or crater floors (Figures 6.10–6.12). For long they have been though to be generically independent. New photographic evidence – such as that reproduced on Figure 6.9, secured by Apollo 15 in July 1971 – demonstrates, however, a clear connection to exist between the two. In this particular instance, our photograph shows a hybrid formation – partly a ridge, and partly a rille at opposite ends – the rille having arisen no doubt by subsidence of the ceiling of a ridge. The wrinkle ridges have long been thought to be akin to the terrestrial 'lava tubes'; and if their roof

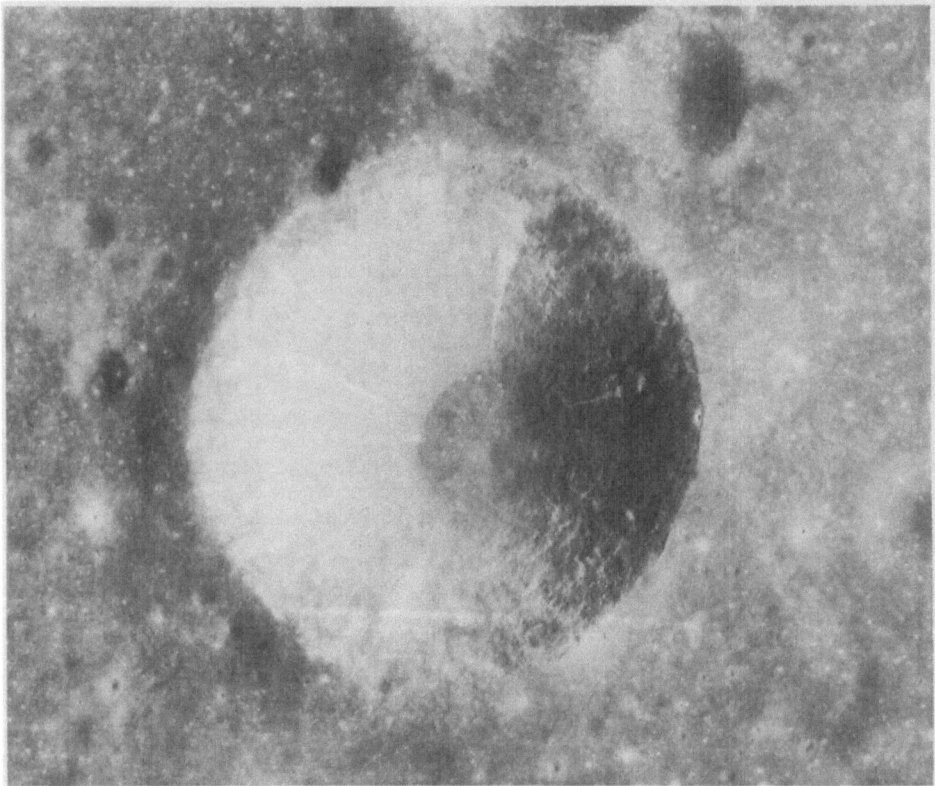

Fig. 6.3.   A crater near Römer in Taurus Mountains north of Mare Tranquillitatis, photographed by the panoramic camera of the Apollo 15 mission in July 1971 from an altitude of 108 km above the lunar surface (NASA official photograph).

Fig. 6.4. The crater Bessel in Mare Serenitatis, photographed by the panoramic camera of the Apollo 15 mission in July 1971, from an altitude of 108 km above the lunar surface (NASA official photograph).

is cracked by mechanical damage (such as can be inflicted by a direct hit of an impinging meteorite; or by cumulative effect of nearby impacts or moonquakes), the outcome may be a rille.

That rilles on the Moon are no dried-up beds of any flow of liquid seems now to be generally acceded, in the face of all evidence which has made such flows extremely unlikely at any time. Instead, the rilles are probably collapsed wrinkle ridges; surface cracks produced by internal stresses; or – for wide formations of this type – just loci of subsidence along pre-existing lines of surface weakness. It is quite possible that, in the case of doubly-terraced formations of this type such as the well-known Schröter's canyon (see Figure 6.8) the inner and outer terracing was produced by different processes at different times. Whether or not the lunar domes – another type of formation of indubitably internal origin (see Figures 6.11 or 6.12), and as characteristic of the lunar landscape as the rilles or wrinkle ridges – are hollow blisters, or structures of more complicated kind, remains as yet unclear.

Another phenomenon which we learned to understand better from the evidence obtained between 1969–1972 concerns the reasons why some parts of the lunar surface appear to us to be brighter than others (i.e., possess different albedo). These reasons are partly chemical (difference in composition) and partly physical (differ-

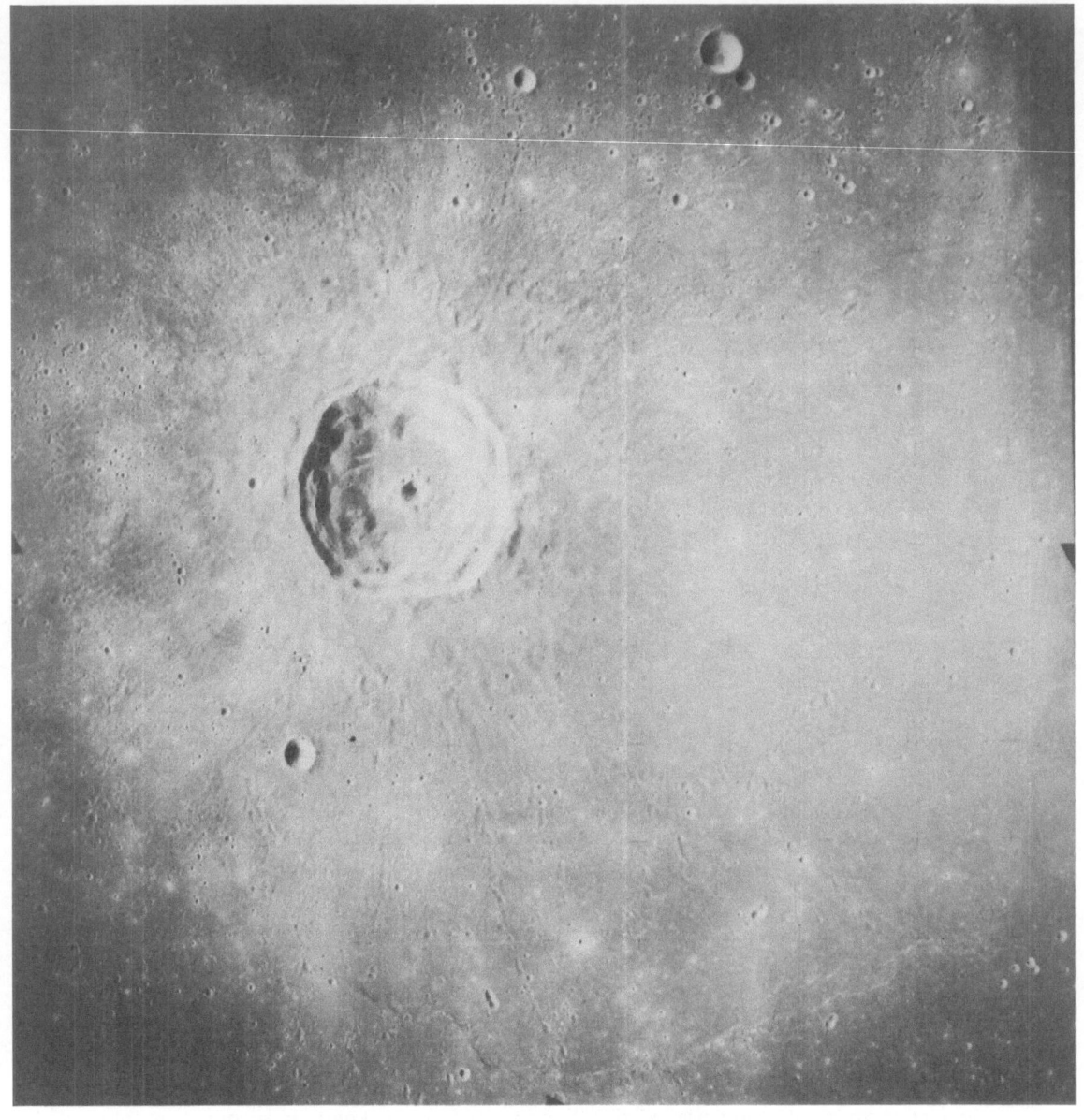

Fig. 6.5. The crater Timocharis in Mare Imbrium, photographed by the metric camera of the Apollo 15 mission in July 1971 from an altitude of 109 km above the lunar surface. Note its collapsed central peak (NASA official photograph).

ence in structure). The main reason why the continental areas on the Moon are brighter than typical mare ground is compositional (greater prevalence of aluminium compounds in the continental anorthosites – in contrast with the magnesium in the mare basalts); and the same is (at least to some extent) true of the 'bright rays' and aprons splashed out of the impact craters of more recent origin – such as one shown on the accompanying Figure 6.13.

However, the latter photograph discloses also the existence of darker material superimposed on the bright apron; and for material forthcoming from the same

Fig. 6.6. A photograph of the Palus Putredinis on the eastern shores of Mare Imbrium, with the Hadley rille meandering along the foothills of the Lunar Apennines. The bay near the eastern bend of the rille was the landing place of the Apollo 15 mission in July 1971; and Mount Hadley – the highest peak of the Apennines (seen from ground on Figure 2.10) dominates the skyline. Photograph taken with the Apollo 15 metric camera from an altitude of 102 km above the lunar surface (NASA official photograph).

source the difference in their reflectivity can hardly be due to different composition. Moreover, craters are known (cf., e.g., Figure 6.14) where dark material actually predominates in the ejecta. A clue to this phenomenon may have been provided by certain observation pertaining to the individual Apollo landing sites. It has been noticed from orbit that, as the astronauts were going busily about their appointed tasks, the ground around them gradually darkened – no doubt because human activity disturbed the fine structure of the topmost surface layer and interfered with

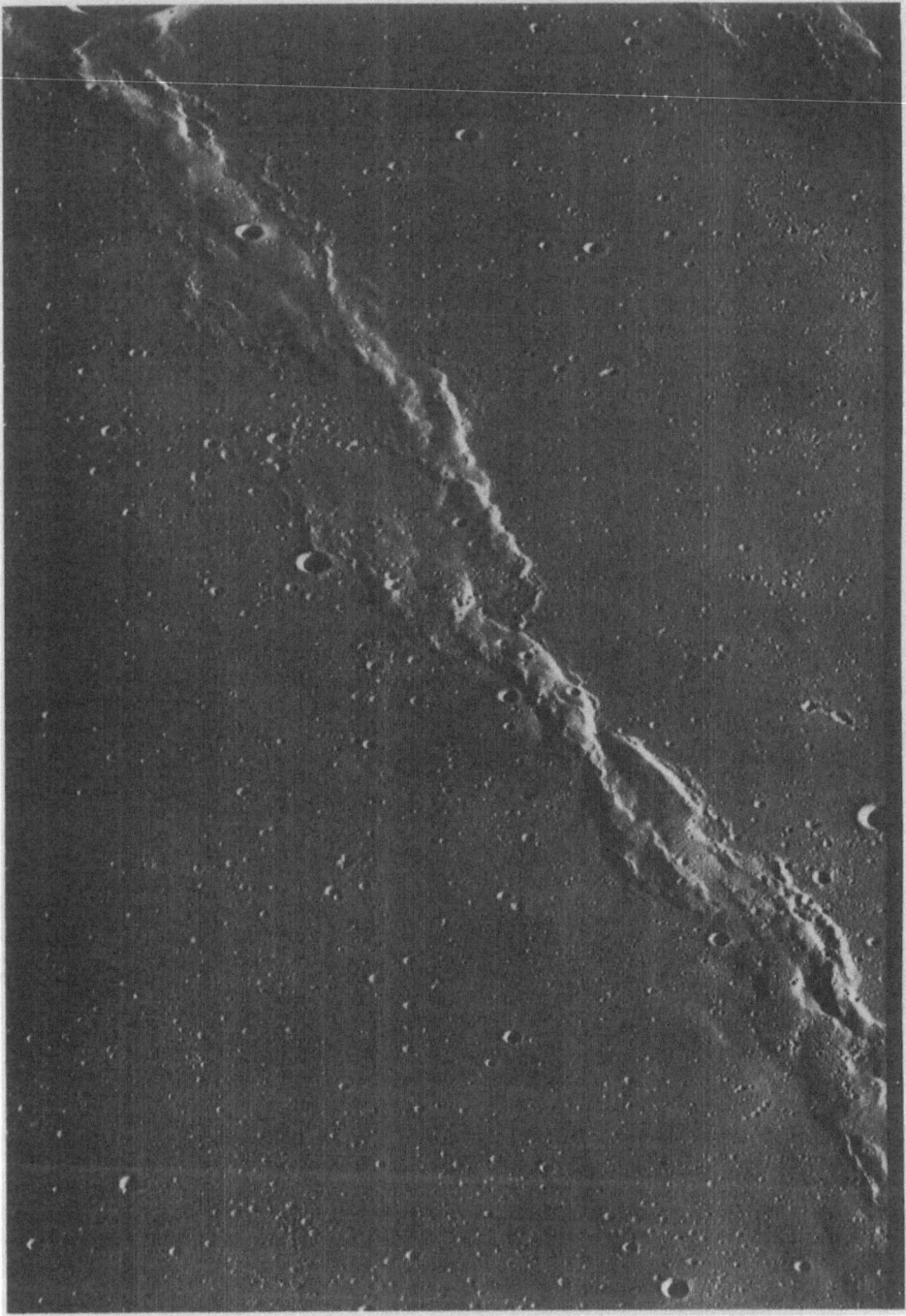

Fig. 6.7.   A part of a wrinkle ridge in Oceanus Procellarum west of the crater Herodotus, as photographed
by the panoramic camera of the Apollo 15 mission in July 1971 from an altitude of 106 km above the
lunar surface. The crater partly seen on top is Lichtenberg B. (NASA official photograph).

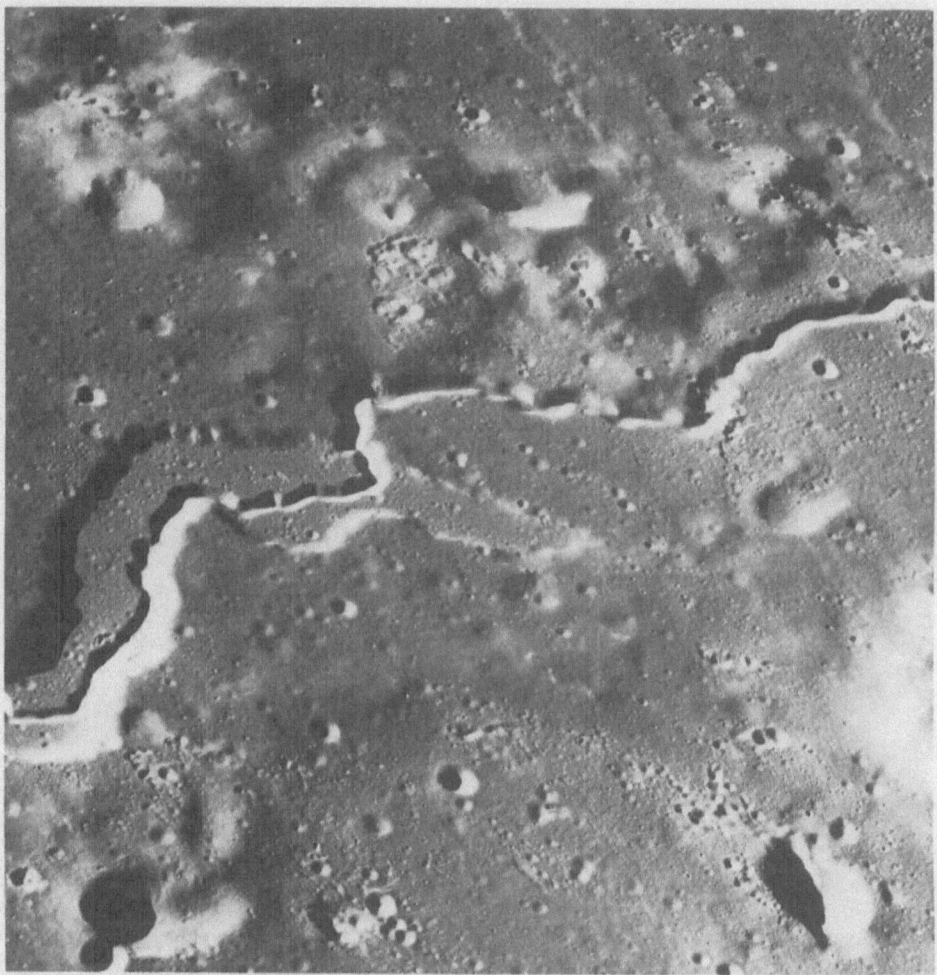

Fig. 6.8. A panoramic photograph of the tail of Schröter's canyon north of the craters Aristarchus and Herodotus (cf. Figure 6.2), showing the double terrace of the feature, taken in the course of the Apollo 15 mission in July 1971 from an altitude of 108 km above the lunar surface (NASA official photograph).

its light-scattering properties. In fact, the more we compress the lunar surface material, the darker it becomes – not because of any change in chemical composition, but because mechanical disturbances have changed its optical characteristics. And what can be caused by human action, can likewise happen as a result of natural processes on a much larger scale. It is, therefore, quite possible that 'dark spots' on the white apron surrounding the crater shown on Figure 6.13, or the dark apron seen on Figure 6.14, can be due (partly or wholly) to albedo changes brought about by mechanical action connected with the origin of the respective formation.

Next, let us turn our attention to another distinctive (albeit macroscopically less spectacular) feature of the lunar surface, unveiled for us already by the U.S. orbiting satellites between 1966–1968, which should enable us to appreciate the effects of

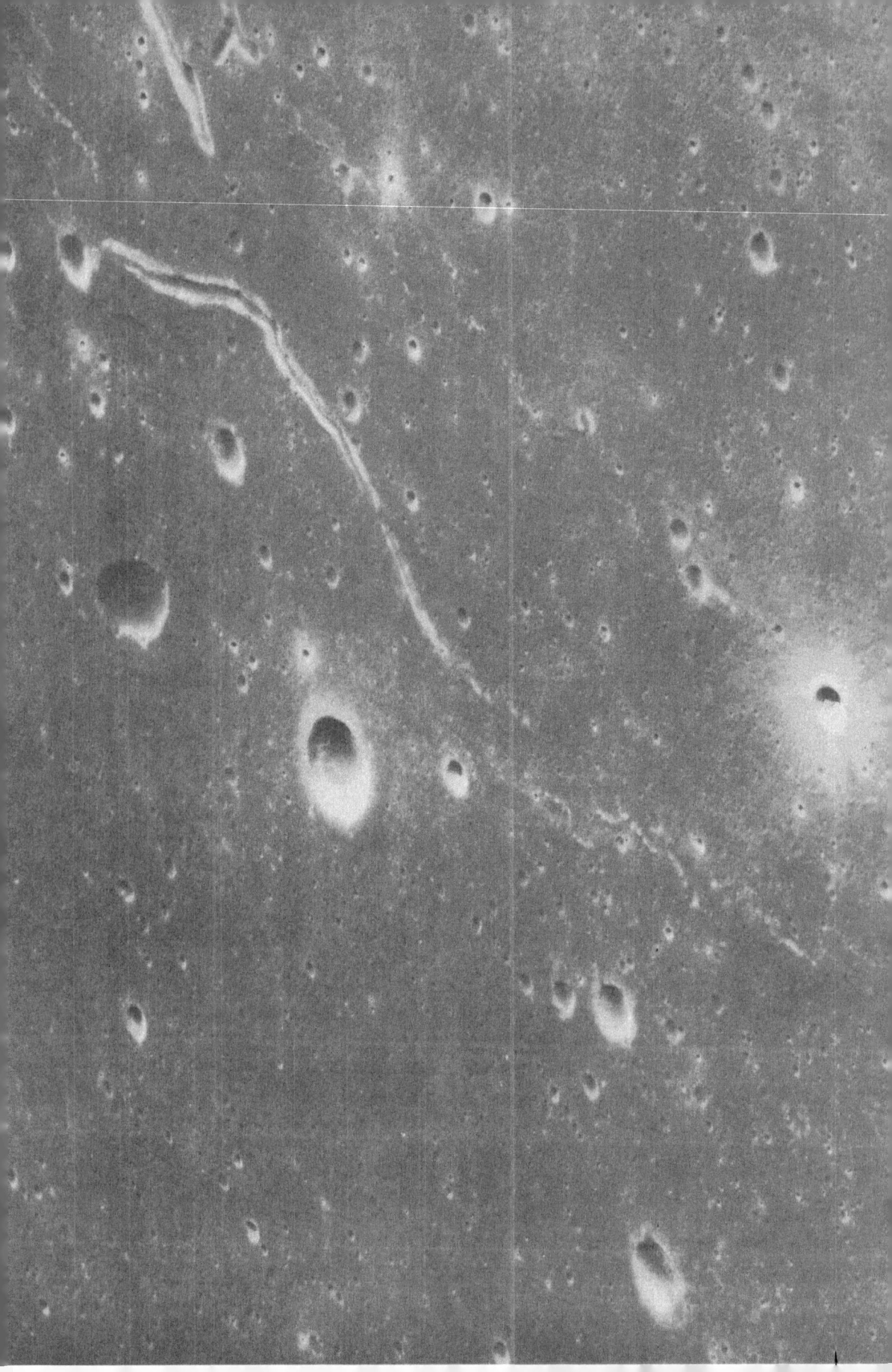

Fig. 6.10.   A part of the floor of the crater Copernicus north of its central peak, taken on 16 August 1967 with the moderate-resolution optics of the Lunar Orbiter 5 from an altitude of 103 km above the lunar surface. The area enclosed in the white square above is the same as reproduced, in high resolution, on the following Figure 6.11; while the lower white square includes the area reproduced in high resolution on Figure 6.15 (NASA-Langley photograph).

←

Fig. 6.9.   A wrinkle ridge half-collapsed into a rille in the eastern part of Mare Serenitatis, and photographed by the panoramic camera of Apollo 15 mission in July 1971 from an altitude of 106 km above the lunar surface (NASA official photograph).

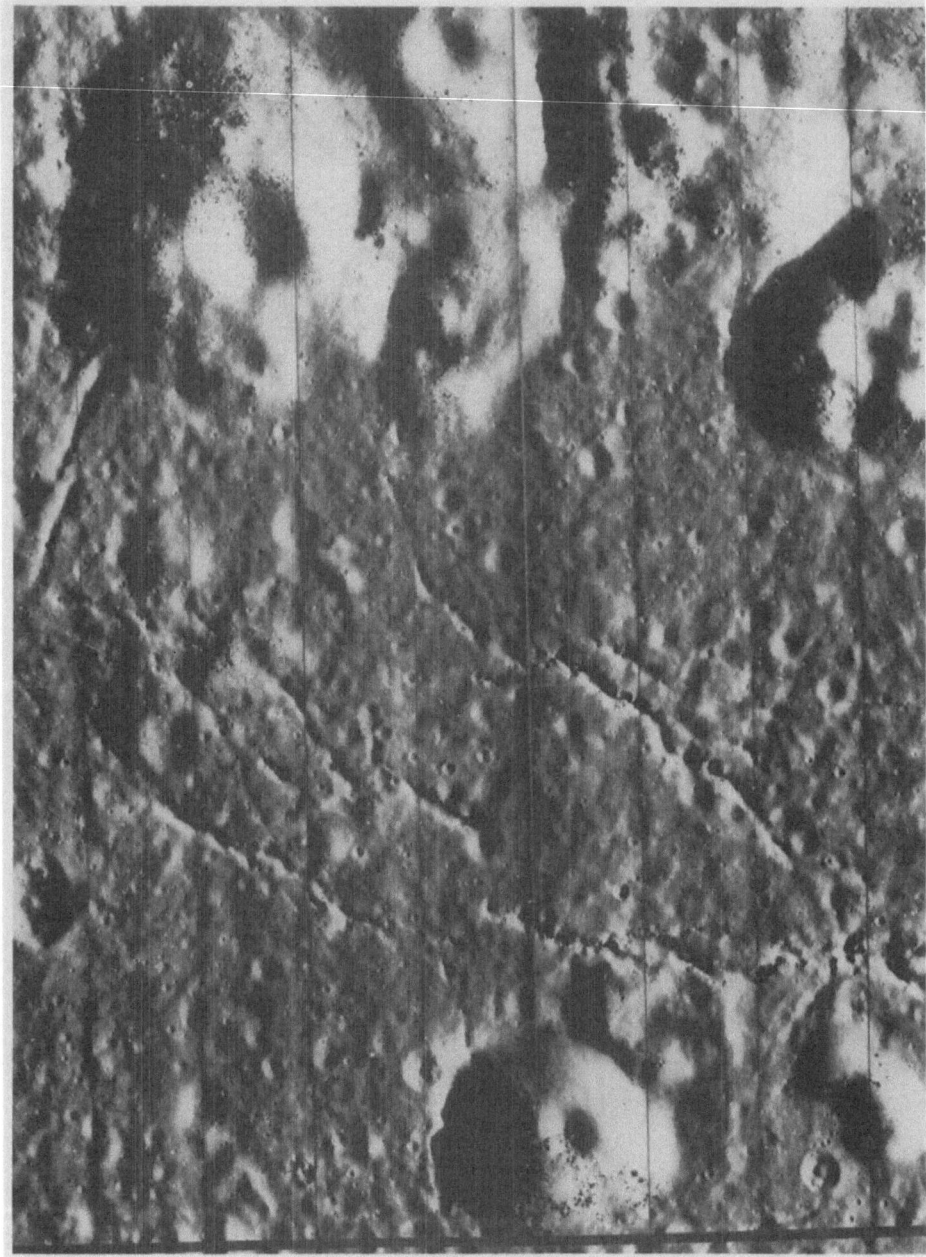

Fig. 6.11.   A lunar dome at a close range – a formation of the same type as that seen north of the Cauchy cleft on Figure 2.14 – photographed by the high-resolution lens of Lunar Orbiter 5 on 16 August 1967 from an altitude of 103 km above the lunar surface. The dome seen near the bottom of the frame is situated (with several others) on the floor of the crater Copernicus; and the boulders obviously thrown out of its summit orifice are only a few meters in size (NASA-Langley photograph).

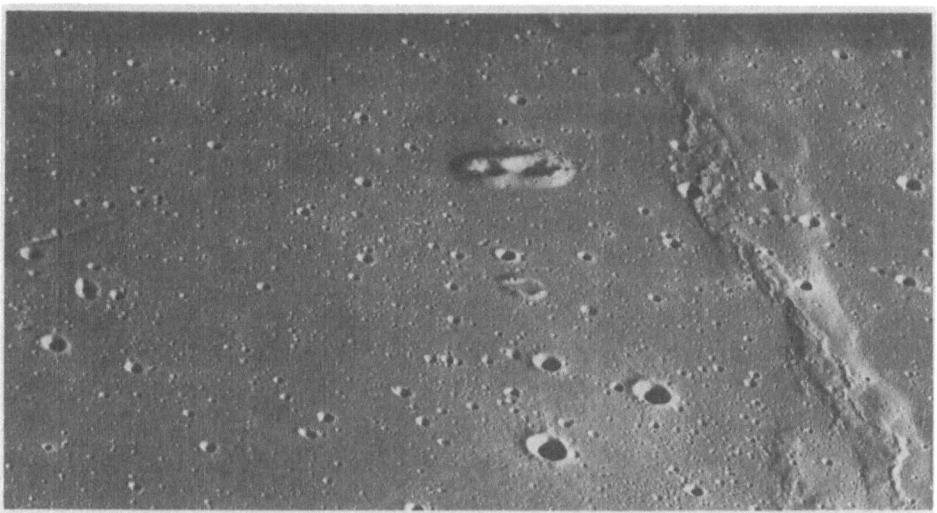

Fig. 6.12. A half-collapsed dome west of the Schröter canyon in Oceanus Procellarum, photographed by the panoramic camera of Apollo 15 mission in July 1971 from an altitude of 107.5 km above the lunar surface (NASA official photograph).

secular *erosion* which have been shaping up the lunar landscape over long intervals of time. In order to do so, let us inspect at a close range (say) the interior of the crater Copernicus shown already before on Figures 1.27 or 6.10. The accompanying Figure 6.15 reproduces a high-resolution photograph of the slopes of one of the hills constituting the central mountain of this crater, taken by the U.S. Lunar Orbiter 5 in August 1967, and showing details 1–2 m in size on the lunar ground.

The particular feature of Figure 6.15 to which we wish to draw attention is the presence, in the field, of a great many boulders – mostly 2–20 m in size – on the slopes of the central hill; and their accumulation at the foot of the mountain – in contrast with their scarcity on the flat crater floor. This is, moreover, no specific characteristic of this particular region; wherever the spacecraft photography recorded for us inclined slopes on the Moon at a sufficient resolution – such as a part of Hadley's rille shown on Figure 6.16 – boulders of comparable size are seen to have detached themselves from the inclined slopes of the rille and rolled down to its bottom. In soft ground, such rolling boulders are apt to leave snakelike tracks behind them to indicate the direction of motion (see Figure 6.17); or distinct traces of avalanches down the inward-sloping crater walls can be identified – such as the one shown on Figure 6.18 in crater Aristarchus – leaving no room for doubt as to the mechanism which produces them. Boulders of the size shown on these photographs (and doubtless many more below their limit of resolution) constitute an essential ingredient of the lunar ground. In this ground, they may be intermingled with finer-grain soil, from which they can get pried loose by mechanical action (moonquakes?); and, propelled by gravity, roll down the slope into the valley.

Preliminary statistics of size-frequency distribution of lunar boulders, based on

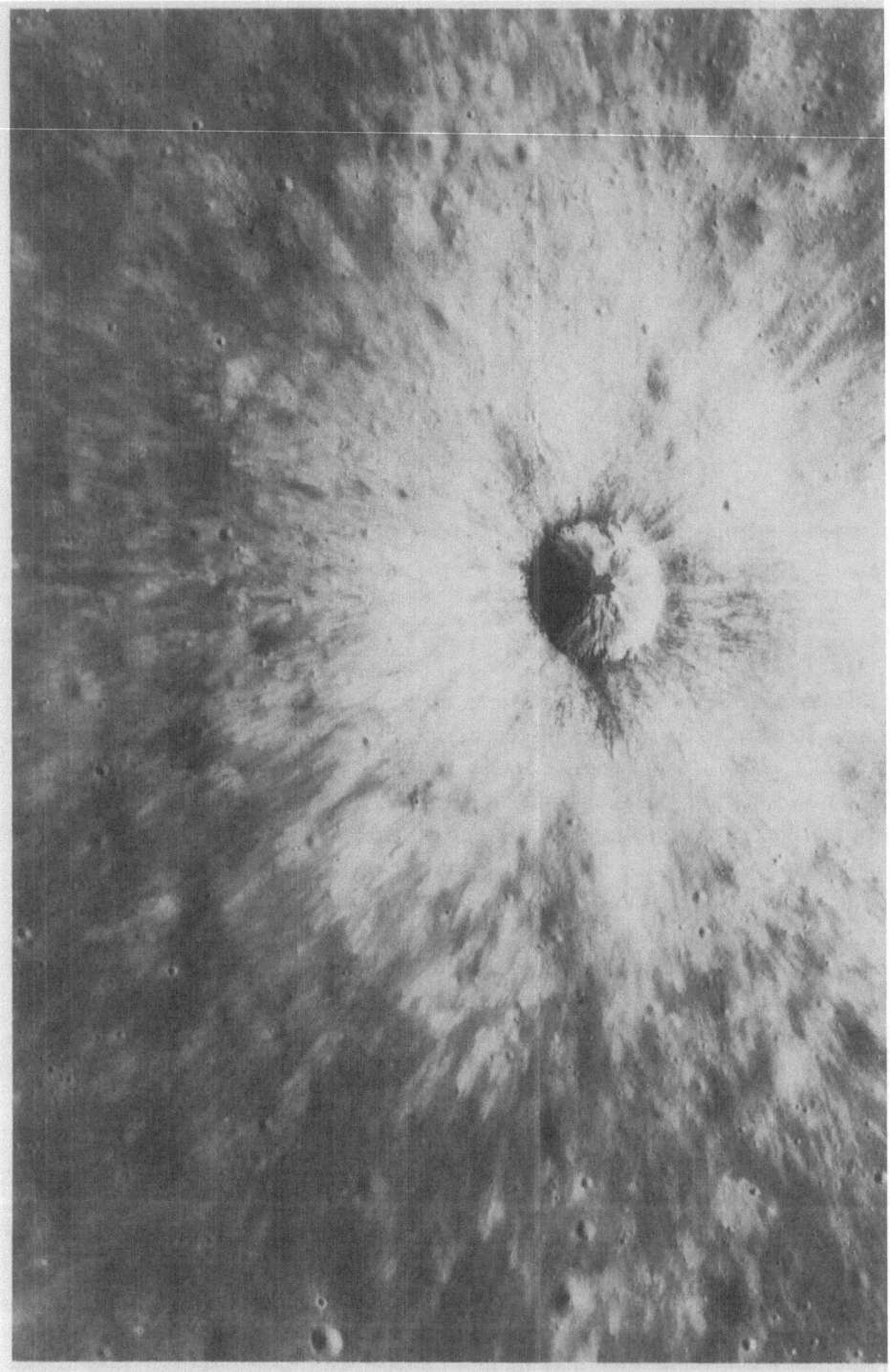

Fig. 6.13. An unnamed splash crater on the far side of the Moon (near the crater Gagarin), showing a conspicuous bright apron of ejecta interspersed with dark streaks, and photographed by the panoramic camera of the Apollo 15 mission in July 1971 from an altitude of 81 km above the lunar surface (by courtesy of NASA and ACIC).

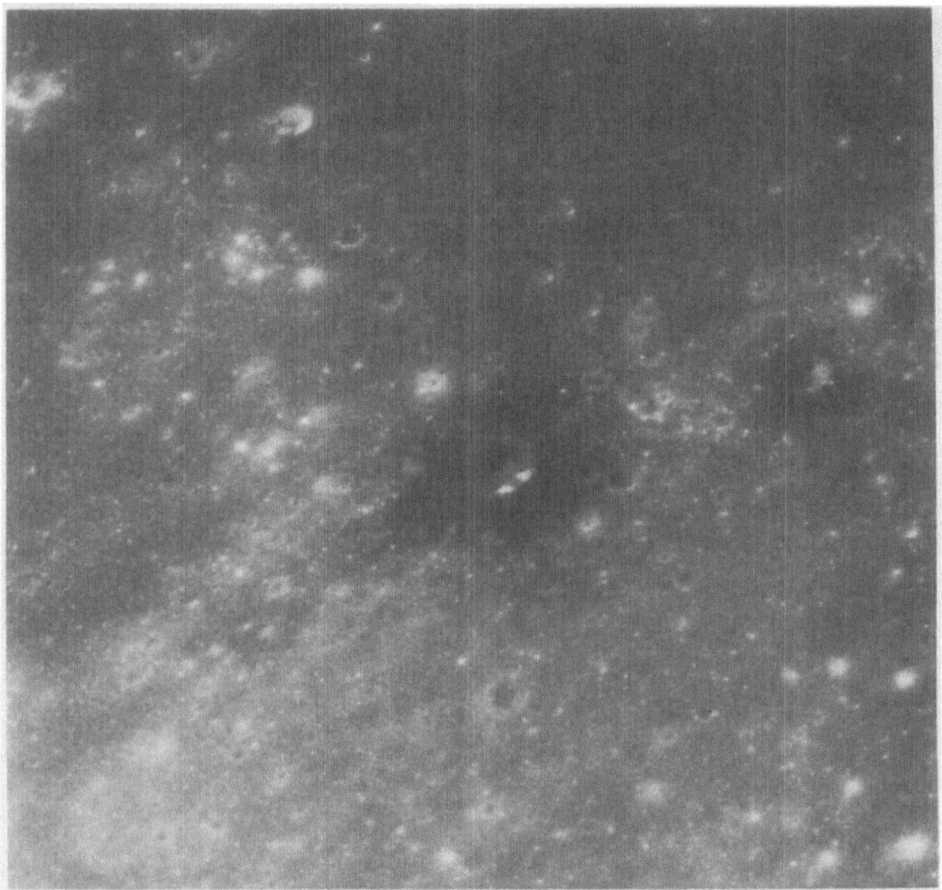

Fig. 6.14. Two dark halo craters (unnamed) in the Haemus mountains, formerly suspected to be of recent volcanic origin. Photograph taken with Apollo 15 panoramic camera from an altitude of 117 km above the lunar surface (by courtesy of NASA and ACIC).

high-resolution photographs of different parts of the lunar surface, have disclosed that a large majority of individual boulders of this type are less than 10 m in size, and very seldom – if ever – exceed 20 m. The well-known 'spires' photographed by Orbiter 2 on 21 November 1966 in Mare Vaporum (see Figure 6.19) constitute boulders the largest of which is about 15×20 m in size – its height above ground having been determined from the length of its shadow cast on the surrounding landscape, while the Sun was approximately 11° above the horizon. The fairly well-defined upper limit on the size of the boulders is probably due to a finite strength of the rocks; for no larger boulders could be made, or transported, in one piece.

High-resolution photographs of the floor of the crater Copernicus have disclosed other interesting and characteristic features resembling those seen on Figure 6.11. In order to demonstrate these, we reproduce on Figure 6.20 a group of small domes which abound in this part of the lunar surface – small hills 200–300 m across and generally less than 40–50 m in height. The characteristic feature of these domes are, however, the fields of one 'gendarmes' covering the brow of these hills, giving them the characteristics of 'hedgehogs'. More than thirty such 'hedgehogs' hills can be located on the floor of Copernicus, one of which is shown in greater detail in Figure 6.21. Their characteristics are all alike: they constitute boulder fields on the tops of low domes, clustering where surface curvature is smallest, and well-nigh absent on the slopes or near the foot of these hills as well as on the flatlands between them.

The significance of such formations is also possible to account for by gravity-propelled landslides. As in the case of the larger hills shown in Figure 6.15, the material of the small domes consists of a mixture of boulders with smaller debris. The stony 'gendarmes' on top of hedgehog-like domes represent probably boulders more securely perched on shallow hills to withstand processes which may send smaller debris intermingled with them sliding downhill, and thus become denuded in time.

That loose boulders on the Moon do run downhill propelled by gravity there is no doubt; for in doing so they leave behind tracks in soft ground which have been photographed in large numbers (cf., e.g., Figure 6.17). Gravity fails, however, to ac-

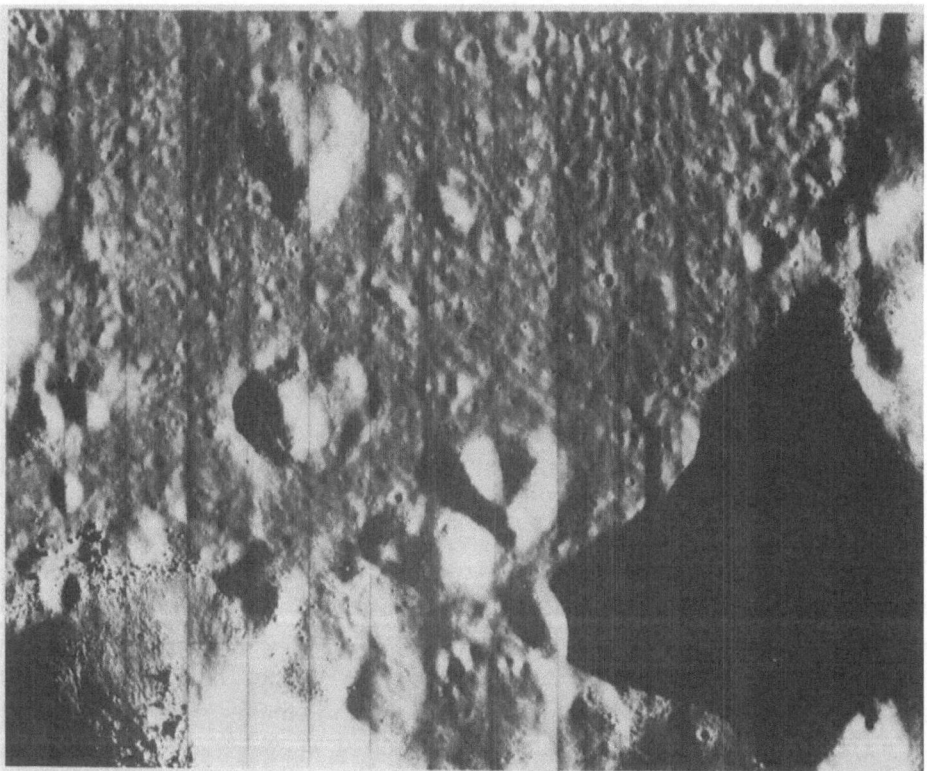

Fig. 6.15. A Lunar Orbiter photograph of the slopes of the hills constituting the 'central mountain' of the crater Copernicus (cf. also Figure 1.27), taken with the spacecraft's high-resolution optics on 16 August 1967 from an altitude of 103 km above the lunar surface. Note the boulders on the slopes and at the foot of the hill on the left (NASA-Langley photograph).

Fig. 6.16. A high-resolution photograph of Hadley's rille (cf. also Figure 6.6), taken by the Lunar Orbiter 5 on 15 August 1967 from an altitude of 104 km above the lunar surface. Note the boulders at the trough of the valley which rolled down the clopes to the bottom (NASA-Langley photograph).

count for the presence of boulders (and boulder fields) which have been discovered also in complete flatlands – of boulders which under no condition could possibly have rolled down to their resting places from anywhere in the proximity. Examples of such boulders photographed *in situ* by the astronauts of the Apollo 15–17 missions have already been seen on Figure 2.9; and it is to these – much more enigmatic – stray boulders that we now wish to turn our attention.

In order to illustrate further our predicament, let us exhibit, on Figure 6.22, a photograph of approximately half a km square of lunar ground in the region of Mare Tranquillitatis, taken with the high-resolution lens of Lunar Orbiter 2 on 16 November 1966. This photograph discloses the presence, in this region, of boulder fields of quasi-circular pattern (reminiscent somewhat of the terrestrial 'megalithic rings'); the individual boulders being 1–10 m in size. Whence did such boulders come? Since there are no hills within miles from which these rocks could have rolled down, the only possibility would seem to be their transfer by impacts.

A possibility that any stone discernible on Figure 6.22 could represent a primary impacting body can, of course, be dismissed out of hand; for such bodies are known to strike the Moon with velocities of many kilometres per second; and no stone impinging on solid surface with such speed would have the least chance to come to rest on the exposed lunar surface for us to see it. Such cosmic intruders would get buried underground at depths equal to many times their original dimensions, where they would probably get vaporized (by a conversion of a large part of its initial kinetic energy into heat), or at least completely shattered into small fragments.

It is known, however, that all major primary impacts capable of giving rise to impact craters of the size of Copernicus or Tycho eject in this process 10^2–10^3 km^3 of lunar material, a large part of which will fall back on the lunar surface almost all over the Moon, along suborbital trajectories characterized by terminal velocities of the order of 1 km s^{-1} (cf., e.g., Kopal, 1966). Local explosions caused by primary impacts can eject – and accelerate to suborbital speeds – boulders of the size we see on Figure 6.22. But – and this is essential – they cannot land them gently enough to survive impact in one piece, and leave them in exposed positions as we see them today.

Fig. 6.17. A high-resolution Orbiter 5 photograph of the lunar ground in the neighbourhood of the crater Vitello, taken on 17 August 1967 from an altitude of 167 km. The photograph shows the sinuous tracks, in the soft ground, left behind two boulders resting at the end of each track.
(NASA-Langley photograph).

Fig. 6.18.   An avalanche of boulders which rolled down the inner slopes of the north-western section of the crater Aristarchus (see Figure 6.2), the trace of which was photographed by the Lunar Orbiter 5 on 18 August 1967 from an altitude of 130 km above the lunar surface (NASA-Langley photograph).

Fig. 6.19.   An area of approximately 250 × 170 m of the lunar surface near the Ariadeus rille ($\lambda = 15°3$ E, $\beta = 4°$ N), as recorded by the high-resolution lens of the Lunar Orbiter 2 on 7 November 1966 from an altitude of 49 km above the lunar surface, showing photograph of some of the largest boulders (casting long shadows) on the Moon. Size of the white-cross mark on the print corresponds to approximately 8 × 8 m on the lunar surface (NASA-Langley official photograph).

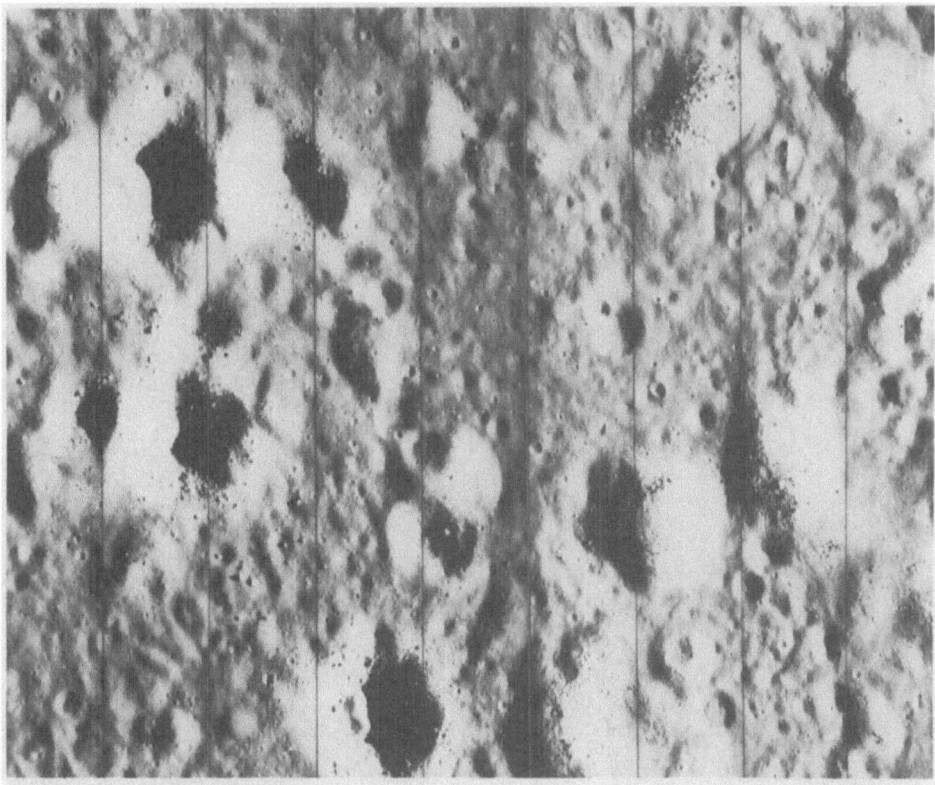

Fig. 6.20. A group of small domes on the floor of the crater Copernicus (north-west of its central mountain), with fields of boulders perched up on their tops, as photographed by the high-resolution lens of
the Lunar Orbiter 5 on 16 August 1967 from an altitude of 103 km above the lunar surface (NASA-
Langley photograph).

In order to demonstrate this in more specific terms, consider the bearing strength
of the mare ground of the lunar surface as revealed already in 1966–1967 by the soft-
landing Surveyors. The area of each one of the three footpads on which these spacecraft came to their rest was close to 700 cm^2, supporting at touchdown a weight* of
approximately 300 kg. Therefore, for a total support area of 3×700 cm^2, the spacecraft would have exerted on the lunar surface a static load of 3×10^5 g/3×700 cm^2 =
143 cm^{-2}. In actual fact, the Surveyors landed on the Moon with a terminal velocity
close to $3\frac{1}{2}$ m s^{-1}; and the dynamical load at touchdown was sufficient to depress
the ground by (typically) about 10 cm below the level of the adjacent surface.

Now a cubic rock of side a and density ϱ will weigh ϱa^3, exerting a static load
per unit area equal to ϱa. For silicate (basaltic) rocks of the kind constituting the
lunar maria, the density ϱ is close to 3 g cm^{-3}. If so, however, a cubical basaltic rock
should exert the same static pressure of 143 g cm^{-2} as the Surveyor's legs if its side
$a = 48$ cm; and should this rock have landed on the lunar surface with the terminal

---

* In the terrestrial gravity field.

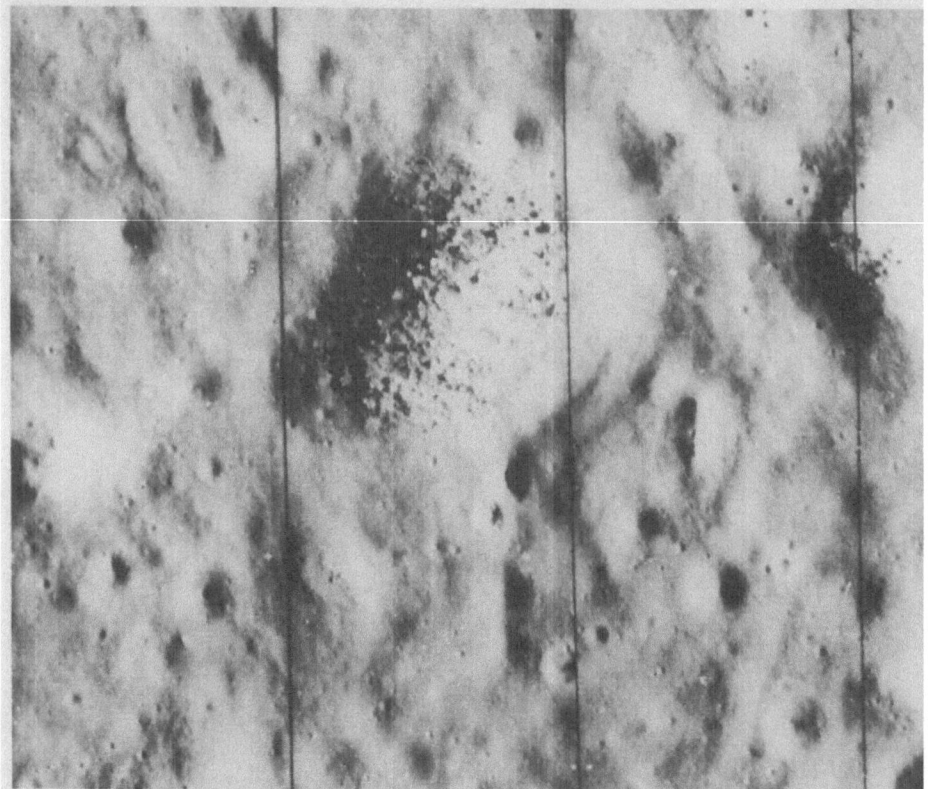

Fig. 6.21.   An enlargement of the structure seen in the upper right corner of the field covered by Figure 6.20, showing the 'hedgehog' structure of one particular dome in this locality. The resolving power of this telephoto frame corresponds to approximately 1 m on the lunar surface (NASA-Langley photograph).

velocity of the Surveyors (3–4 m s^{-1}), it should have penetrated it down to a depth equal to 21% of the length of its side.

The dynamical load exerted per unit area is known to increase (for moderate loads) approximately with the square of the velocity of impact. If so, however, our rock would have buried itself completely underground for impact speed around 10 m s^{-1}; and for velocities of the order of 100–1000 m s^{-1} (typical of secondary or tertiary impacts) the impinging missiles would come to their underground rest at a depth equal to many times their actual dimensions. This depth could, of course, be lessened if the target area were to consist of harder ground. But in such a case no solid rock could withstand impact shock without getting shattered into pieces too small to be resolved on the photographs of the Orbiters or the Apollos taken from the orbit.

In other words, according to any plausible theory which could be formed in advance, such rocks or boulders as the reader sees on photographs reproduced on Figures 6.19 or 6.22 should really not be there! For boulders occurring in flat marial regions of the lunar surface, far from any conspicuous mountains, could scarcely have arrived at their present locations otherwise than by impact along some ballistic

Fig. 6.22. A field of 'megalithic rings' of lunar boulders (1–10 m in size) in Mare Tranquillitatis ($\lambda =$ 34°42 E and $\beta = 2°81$ N), recorded by the high-resolution lens of the Lunar Orbiter 2 on 7 November 1966 from an altitude of 45 km. The size of the field recorded on the photograph is approximately 450 × 360 m (NASA-Langley photograph).

trajectory. If the target ground was soft (as indicated by all the soft-landers), the rock should have buried itself underground well out of sight of the external observer; if, perchance, the target ground was hard, no silicate rocks moving at 100–1000 m s^{-1} could have survived the impact as boulders 1–10 m in size.

What is the way out of this perplexing dilemma? In an effort to find it, let us recapitulate the facts bearing on the case which appear to be observationally well-founded, or at least probable, on grounds of more circumstantial evidence.

(a) Boulders 1–10 m in size are shown by high-resolution photographs of the Moon to occur individually or in groups on many parts of lunar mare ground which are far from any cliffs or conspicuous mountains; these boulders consist probably of basaltic material, of density close to 3 g cm^{-3}.

(b) In the absence of any mountainous ground in the vicinity, these boulders have probably arrived at their present localities by ejection from other parts of the Moon, along suborbital trajectories characterized by secondary impact velocities of the order of 100–1000 m s^{-1}.

(c) The target mare ground possesses a bearing strength of the order of 10^5–10^6 dyne cm^{-2}, as indicated by the soft-landers, down to a depth of at least a few metres (i.e., far beyond the reach of the instrumentation of the soft-landers); for otherwise impacting basaltic boulders would get shattered into pieces too small to be resolvable on our photographs.

(d) If soft ground is to cushion the impacts so as to enable rocks 1–10 m in size to survive in one piece, these rocks must be able to penetrate into the ground by at least a few times their original size. In other words, rocks transported by secondary impacts should disappear out of sight in the maria, and only shallow craters or depressions should mark their burial place immediately after the event.

(e) If, therefore, such rocks eventually emerge again on the surface to become visible to an external observer, it follows that they must have subsequently been *exhumed* by some kind of *erosive process* that gradually removes the debris in which the rocks became originally imbedded. This seems to be the only possibility for large rocks to be seen now in one piece in the midst of the relatively soft mare ground, far away from any exposed cliffs or mountains.

What kind of erosion could be invoked to accomplish this task? It goes without saying that such processes as micrometeoritic bombardment or sputtering caused by solar wind, which endow the lunar micro-relief with its peculiar photometric properties, are completely ineffective in this connection. However, another process is known to exist which can better accomplish the purpose, and this is the secular *cumulative effect of moonquakes*.

That seismic quakes are bound to occur on the Moon – due to external as well as internal regions – was anticipated with confidence before 1969; and ever since the first seismograph was installed on the Moon by the Apollo 11 mission in July 1969, this anticipation has been amply confirmed by actual evidence. The main results of this work have already been described in the preceding chapter of this book. We may recall, in particular, the fact that the yearly energy release, by seismic waves, throughout the lunar interior amounts to 10^{15} erg now; and may have been very much larger in the past when the cosmic bombardment of the lunar surface was very much more lively.

One consequence entailed by this intermittent but long-drawn cosmic bombardment of the Moon by external bodies must have been a gradual break-up and shattering of its surface rocks (through the action of shock and seismic waves emanating from shallow epicentres of individual impact points) down to a considerable depth. That this has been so is no mere theoretical surmise, but an actual fact discovered by an analysis of the observed returns of radar echoes from the lunar surface, going back some twenty years before the advent of soft-landing spacecraft.

Ever since the first echoes of radar pulses sent out to the Moon have been recorded in 1946, it was found that these echoes were considerably weaker than they should have been for a globe of lunar distance and size, consisting of solid rocks. In order to reconcile theory with the observations it was necessary to assume that the effective dielectric constant of the lunar surface material was about three times smaller than that of common silicate rocks. Between 1946–1966 the same conclusion was arrived at by experiments with radar pulses at very different wavelengths, and reflected from surface layers of very different depth (equal, in most cases, to several multiples of the wavelength). It was not until pulses at decametre waves were first reflected from the Moon (cf., e.g., Davis and Rohlfs, 1964) that the strength of the echoes began to point to dielectric constants approaching those of solid rocks; and these echoes (for pulses corresponding to wave-lengths $\lambda > 15$ m) returned from sub-surface depths between 50–100 m.

On the basis of the mean density of 3.34 g cm^{-3} of the lunar globe, it has long been surmised that the Moon consists essentially of silicate rocks similar to those common in the Earth's mantle; and an analysis of the rocks imported from different parts of the lunar surface by the Apollo and Luna missions of 1969–1972 have confirmed this surmise. Under these conditions, the low effective dielectric constants inferred for the lunar surface from radar echoes (and also from the degree of polarization of its thermal emission in the microwave domain) could be reconciled with their now known composition only by assuming a *low volumetric concentration* of the material – i.e., an assumption that the reflecting layers are not solid, but consist of an accumulation of rubble and loose debris with a considerable fraction of empty space between individual grains or stones. The same conclusion has also been independently arrived at from studies of thermal-conduction properties of the surface and, more recently, by direct samplings of the lunar surface material by the soft-landers.

However, none of these methods prossesses the penetrating power of radar echoes at decametre waves; and these have indicated that diminished volumetric concentration of lunar surface material extends down to a depth of 50–100 m on the average – a long way beyond the reach of any sampling device now in operation or contemplated. As the radar echoes emerging at decametre waves possess low angular resolution (with the optics employed, the Moon would in their light appear as a single picture-point), low volumetric concentration of the surface material deduced from them to a depth of 50–100 m represents only a global average which can be exceeded in some places, and not attained in others. However, that the Moon borders on space through a brecciated lithosphere – i.e., a shell of broken material (of relatively low volumetric concentration), many metres in depth can now be accepted, not only as a theoretical expectation, but also as an observed fact (cf. Figure 6.23), and as a global characteristic of our satellite, to which the geologists refer as a 'regolith'.

The cause of this mechanical damage – namely, primary cosmic bombardment – is, to be sure, not operative on the Moon alone, but also on any other celestial body equally exposed to it. In particular, our Earth must have received a comparable bombardment, per unit area, in the same length of time; since for cosmic bodies causing greatest mechanical damage – intruders with masses greater than 10^6 g – our atmosphere provides next to no protection. However, the combined action of air and water would obliterate any wounds caused by cosmic bombardment very much more rapidly than these are inflicted, so that our lithosphere shows no evidence of a global fragmented layer. The fact that our Moon does exhibit evidence of such a layer discloses that the 'healing agents' – air or water – are not only absent on the Moon now (as we know abundantly also from other sources of information), but that they probably were of no importance over the long astronomical past of our satellite.

But having made a case for a global layer of fragmented material to extend to a depth of many metres over most of the lunar surface, let us inquire whether or not the existence of such a layer can help us to understand the presence of boulders in exposed regions of the lunar flatlands, as shown in Figures 6.19 or 6.22. The answer is indeed in the affirmative; for if such boulders form a constituent part of the lunar regolith – having been imbedded in it by secondary impacts (or produced by shattering effects of primary impacts *in situ*) – they could be made to surface by cumulative shake-up action caused by individual moonquakes of external (impact) as well as internal origin. For it is well known from soil mechanics that one of the most effective ways to sort out loose debris or gravelly material of the same density but differing in size is to administer the mixture a mechanical shake-up – in the course of which large particles will eventually emerge on the surface, being supported by smaller particles which tend to slip through and subside towards the bottom.

The presence of individual boulders on the level parts of the lunar surface – in exposed positions which the static bearing strength of the surface is barely able to support – demonstrates, according to our view, the outcome of a gradual mechanical 'erosion' of the kind described in the preceding paragraphs, in the course of which *boulders* or large stones *imbedded in loose material are gradually lifted* ('litho-exhumed') *by shake-up produced by moonquakes*, eventually to make their way to the surface. According to this view, the 'megalithic rings' of boulders shown in Figure 6.22 could represent the debris of a much larger rock which once fell there and split up into pieces underground. What we see now are remnants of this event, in the form of fractured remains which have gradually made their way to the surface by intermittent action of mechanical erosion.

Fig. 6.23. A telephoto taken by the astronauts of the Apollo 15 mission of the western downward-sloping wall of the Hadley rille (cf. Figure 6.16), showing the stony structure of the topmost layer of the lunar regolith in this region. The edge of the rille as shown in this view is approximately 150 m in length; and the sizes of the large individual boulders run between 5–8 m (NASA official photograph).

The views developed in this chapter bring into common focus several known but unrelated facts and processes, the participation of which is essential to produce mechanical erosion of which stray boulders on the lunar surface may be an external manifestation. They include: (a) the existence of a loose layer of surface debris over-lying compact rocks to a depth of many metres (i.e., large in comparison with the size of the boulders); (b) the production mechanism of this loose layer (i.e., cosmic bombardment) generates also moonquakes that tend to bring about a gradual sub-sidence of the loose layers, in the course of which large particles are gradually brought to emerge on the surface; and (c) the absence of lithification or other processes which could make the loose layer to lithify before the sorting of particles by moonquake-produced shake-ups could have been accomplished.

The question can be asked: is there any evidence for the operation of a similar process on the Earth? The answer is in the negative; and must necessarily be so on any planet whose surface is at the mercy of much more powerful erosion produced by air and water. To be sure, impacts sufficiently powerful to produce craters 1 km in size or greater pay no attention to our atmosphere on entry, and shatter rocks around the point of impact in much the same way as they do on the Moon. But once the terrestrial damage has been inflicted, the scar will heal quickly (on a geologic time scale) by combined effect of air and water; so that (say) a million years later traces of the original catastrophy could be identified only by a careful investigation *in situ*.

A million years is a very short time in comparison with the average time-interval that is likely to elapse between two successive impacts, capable of producing craters of this size, in any one place. This implies, in turn, that scars ('astroblemes') so produced will heal on the Earth much more rapidly than they are inflicted; so that there is no chance for any damage caused by them to accumulate. It is only on planetary globes as small as that of the Moon – lacking completely any air or water – that the much slower litho-erosive processes described in the latter part of this chapter have any chance to show their hand; and we have reasons to believe that we identified them.

References

Baldwin, R. B.: 1949, *The Face of the Moon*, Univ. of Chicago Press, Chicago.
Davis, W. M.: 1926, 'Bibliographical Memoir of Grove Karl Gilbert, 1843–1918', *Mem. U.S. Nat. Acad. Sci.* **21**, 303.
Davis, J. R. and Rohlfs, D. C.: 1964, *J. Geophys. Res.* **69**, 3257.
Kopal, Z.: 1966, *Icarus* **5**, 201.
Kopal, Z.: 1969, in *Space Research* IX, North-Holland Publ. Co., Amsterdam, pp. 657–677.
Spurr, J. E.: 1944–49, *Geology Applied to Selenology*, vols. 1–4, Science Press, Lancaster, Penna.

CHAPTER 7

SURFACE STRUCTURE AND CHEMICAL COMPOSITION

After a brief survey of the morphology of the principal types of formations encountered on the lunar surface, and reconnoitred more recently *in situ* by the astronauts
of successive Apollo missions, let us turn now our attention to the physical structure
of the lunar surface and its chemical composition.

1. Surface Structure

Until the advent of soft-landing spacecraft in 1966, all investigations of the structure
of the lunar surface had to rely completely on indirect methods, in which the principal
link between us and the object of our inquiry is the lunar radiation in the full range
of its spectrum. With one quite insignificant exception (i.e., thermal radiation of the
Moon due to the leakage of its internal radiogenic heat) all moonlight derives its
origin from the Sun – whether this be sunlight falling on the Moon directly, or scattered towards it through the intermediary of our Earth ('earthshine'). All this light
must be absorbed or scattered by the lunar surface, in accordance with its local optical properties. A small part of the solar radiation (both electromagnetic and corpuscular) may be absorbed and re-emitted by cascade processes giving rise to fluorescence in the visible part of the spectrum. Energetic corpuscles of the solar wind
may even induce the lunar surface to emit 'bremsstrahlung' in the X-ray domain, a
spectrum of which could disclose the atomic composition of the outermost lunar
crust – just as studies of the luminescent spectra at optical frequencies can provide
some information about the molecular structure of lunar material, and impurities of
its crystal structure.

    The first attempts, by Rossi and his associates, at observing lunar fluorescent X-ray
spectra from balloon altitudes in 1962 were unsuccessful in their main objective; but
they led to an interesting and unexpected byproduct: namely, a discovery of the first
discrete galactic X-ray source (Scorpius X-1) which happened to be close to the Moon
in the sky at the time of observations (cf. Rossi, 1973). The lunar X-rays were too
weak to register at the distance of the Earth; and their actual discovery did not come
till four years later with Luna 12 (Mandelshtam *et al.*, 1968), and later when Adler
and his associates (1972a, b) detected their emission by the lunar surface from the
Apollo 15–16 command modules in orbit around the Moon.

    The results of these experiments will be detailed later in this chapter. For the
present we wish to return to the state of the subject before the advent of spacecraft,
and to summarize the accomplishments of the epoch when the telescopes were still

our principal lines of communication with our satellite. At optical frequencies and in the near infra-red (up to wavelengths of approximately 4–5 μ) the lunar spectrum is dominated by scattered sunlight; while at wavelengths longer than 5 μ – i.e., in the deeper IR or the domain of radiofrequencies – the radiation of our satellite is due almost exclusively to a thermal emission of its globe. The energy sent out as thermal radiation is much greater than that of scattered light; for only about 7% of incident sunlight gets scattered from the lunar surface; the balance being absorbed and re-emitted.

Even as large a balance of incident flux is, however, insufficient to maintain the outermost layer of the lunar surface at a temperature higher than 390 K at the sub-solar point, and less than 80 K late at night. This means that much of the thermal radiation of the Moon (radiating like a black body at these temperatures) is bound to be emitted at wavelengths which are absorbed by our own terrestrial atmosphere. Fortunately, this atmosphere is fairly transparent in the 8–12 μ wavelength window, which should include the region of maximum emissivity of the lunar surface in day-time. The second atmospheric window through which the thermal radiation of the Moon can be observed in the microwave domain of the spectrum (for $\lambda > 1$ mm) lies already so far on the descending branch of the intensity-distribution of a black-body emitter of temperature as low as that of the lunar night time that the energy flux received from it is quite small, but still measurable up to wavelengths of the order of one metre.

The main significance of the measurements of the thermal emission of the lunar globe rests on the fact that, inasmuch as such long-wave radiation originates at an increasing depth below the visible surface, the attenuation of the diurnal heat wave and increase of its phase-lag with increasing wavelength provide us with direct means to establish the absolute value of the thermal conductivity (and of dielectric proper-ties) of the lunar surface down to an appreciable depth.

The emitted and scattered components of lunar radiation can also be distinguished by their different polarization properties; for while the scattered part of moonlight becomes distinctly polarized in the process (and the direction of its plane of polar-ization rotates with the phase), thermal emission in the near IR (though not in the microwave domain) remains essentially unpolarized. The same distinction exists also between the illumination of the Moon by the Sun and the Earth: while the incident sunlight is unpolarized, the earthlight is already partly polarized by the scattering of sunlight in our atmosphere.

These are all passive sources of light provided by nature. In addition, it has proved possible in recent years to reflect from the Moon man-made radar pulses correspond-ing to wavelengths ranging from less than one centimetre to almost twenty metres, and to record with sufficient precision their line profiles modified by the reflecting properties of the lunar surface – a very powerful method of exploration, since long-wave radar pulses penetrate much deeper in the lunar crust than the diurnal heat wave.

The principal results of the measurements of different properties of moonlight and its variation with the phase can be summarized as follows:

(1) The intensity of light scattered at optical frequencies varies so rapidly before and after full Moon (the full phase being approximately 19 times as bright as the first or last quarter, when due regard is paid to the 'opposition effect') as to defy explanation in terms of diffuse reflection from smooth surface of any known natural substance. Moreover, the apparent disk of the Moon exhibits no trace of limb-darkening in visible light. These phenomena reveal that, at some scale which is large in comparison with wavelength, the lunar surface must become extremely rough, and capable of an extraordinary amount of back-scattering.

(2) The foregoing statement is, moreover, true not only of the global light of the apparent lunar disk, but also of any element of it – be it a part of the continental blocks or maria. Each element attains its maximum brightness at full Moon, re-gardless of its relative position or angular distance from the Moon's centre.

(3) The reflectivity (albedo) of the Moon varies from place to place within the range 0.05–0.18 in yellow light ($\lambda = 0.56 \mu$) – i.e., much less than for most common terrestrial rocks – and increases somewhat with the wavelength. The ratio of the albedo of the brightest and darkest spots optically resolvable on the Moon exceeds, therefore, scarcely a factor 3; while the continental areas are, on the average, not more than 1.8 times as bright as the maria.

(4) The light of the Moon as a whole is distinctly redder than the illuminating sunlight, and becomes more so with increasing phase (i.e., away from the full Moon); but its local colour differs but little from spot to spot. In general, the magnitude of the colour index of any surface detail appears to increase statistically with its albedo.

(5) The scattered moonlight is polarized to the extent of several per cent. At small phase angles ($0 < \alpha < 30°$) the polarization proves to be negative, but it changes sign thereafter and attains a maximum roughly at a phase of 90°; the direction of the plane of polarization being either parallel with, or perpendicular to, the 'equator of illumination'. The phase at which the phase of polarization rotates so by 90° coincides (for most lunar objects) with the quadratures. The actual amount of po-larization is found to increase with diminishing albedo; the maximum polarization of dark maria exceeds 15%.

(6) The intensity distribution of thermal radiation of the Moon in the 8–12 μ domain of the spectrum exhibits distinct limb-darkening; and the corresponding light (or, rather, heating or cooling) curves are of strongly local character – indicating that the lunar surface is greatly diversified in its thermal properties from place to place.

(7) The intensity of thermal radiation of lunar origin in the 1–1000 mm microwave domain of its spectrum, and its variation during lunation (or eclipse), reveals that the effects of a diurnal heat wave penetrate to a depth of barely a foot below the surface, where a constant temperature close to 240 K prevails day and night. More-over, its phase-lag grows with the depth of penetration in such a way as to indicate a far lower coefficient of heat conduction for lunar surface layers than that of any known terrestrial solid rock – a result explainable only on the assumption that lunar

surface material consists of loose rubble or dust, in which heat can flow only through the corners of contact between individual elements of the debris.

(8) The power reflection coefficients deduced from the thermal radiation of our satellite in the infrared as well as microwave domain of its spectrum lead, moreover, to a dielectric constant ε of the lunar crust which is much smaller than that of solid silicate rocks – indicating again a considerable degree of fragmentation of the surface material (the mean volumetric concentration of the material of not more than 40–50%).

(9) The observations of radar echoes at 30–3000 Mc s^{-1} (10–1000 cm wavelength) reflected from the Moon reveal that approximately 50% of the echo power arises as a result of quasi-specular reflection from a small central region of radius about one-tenth of that of the apparent disk of the Moon. The power reflection coefficients resulting from the observed echo strengths disclose that low dielectric constant ε (indicating low volumetric concentration of the material), inferred previously from observed properties of thermal radiation of our satellite, extend down to a depth from which 10-m radar pulses are returned; and this level may be many dozens of metres below the visible surface. It is not till in the domain of decametre waves that the returning echoes point to an increase in the effective value of ε, indicating an increased compression of the material and, eventually, a contact with solid rocks. But the average level at which loose debris gives way to solid rocks would appear to be somewhere between 50–100 m below the visible surface.

(10) The slope of the trailing edge of the specular part of radar echoes at different wavelengths indicates that the reflecting surface continues to be essentially smooth (with average gradient of one in ten or twenty), and covered with objects below the limit of radar resolution to no more than 10% of its area down to almost centimetre wavelengths. It is not till for radar waves characterized by $\lambda > 1$ cm that the surface begins to appear quite rough (cf. Lynn et al., 1964), as anticipated previously from its light-scattering properties at optical frequencies. In certain localities (ray craters, for instance) radar methods have indicated surface roughness on the metre scale (cf., Pettengill and Henry, 1962; Thompson and Dyce, 1966); and such anomalies appear to be related with parallel thermal or albedo anomalies. The lunar surface appears to be no more a uniform radar reflector than it is a uniform light reflector or heat conductor; and local anomalies in these respects are as numerous as they are conspicuous.

This was, in brief, a simplified model of the structure of the lunar surface as it emerged from deductive work based on telescopic astronomical observations before the advent of the space-age; and we recount it here because subsequent space work has not faulted this picture in any essential respect. Needless to say, when spacecraft of different types began to make their contributions, many of the features observed by more traditional methods found their natural explanations confirmed beyond reasonable doubt. Thus it transpired that the thermal anomalies ('hot spots') observed on the lunar surface are caused mainly by a profusion of boulders on the 1–10 m scale, discovered on the high-resolution photographs taken by the U.S. Lunar

Orbiters between 1966–1968 to populate the location of many 'hot spots' on the Moon in large numbers. What produced these boulder fields where they are found constitutes an interesting question (cf. Kopal, 1969) which occupied us already in the preceding chapter. But there is no doubt that each such boulder possesses a very large capacity of heat absorption during daytime, for subsequent re-emission at night (or during lunar eclipses). After the Sun has set, each boulder will act as an independent black-body radiator, operating at a temperature generally 100 or more degrees higher than that of the surrounding landscape; and an overlapping radiation field emitted by such boulders can maintain their environment at a more elevated temperature – giving rise to a 'hot spot'.

On the other hand, a resort to many ingenious refinements of the range-doppler tracking of the Moon at radar frequencies (for their description, cf. *Moon II*, Chapter 21) enabled Pettengill and his associates at MIT Lincoln Laboratory to construct

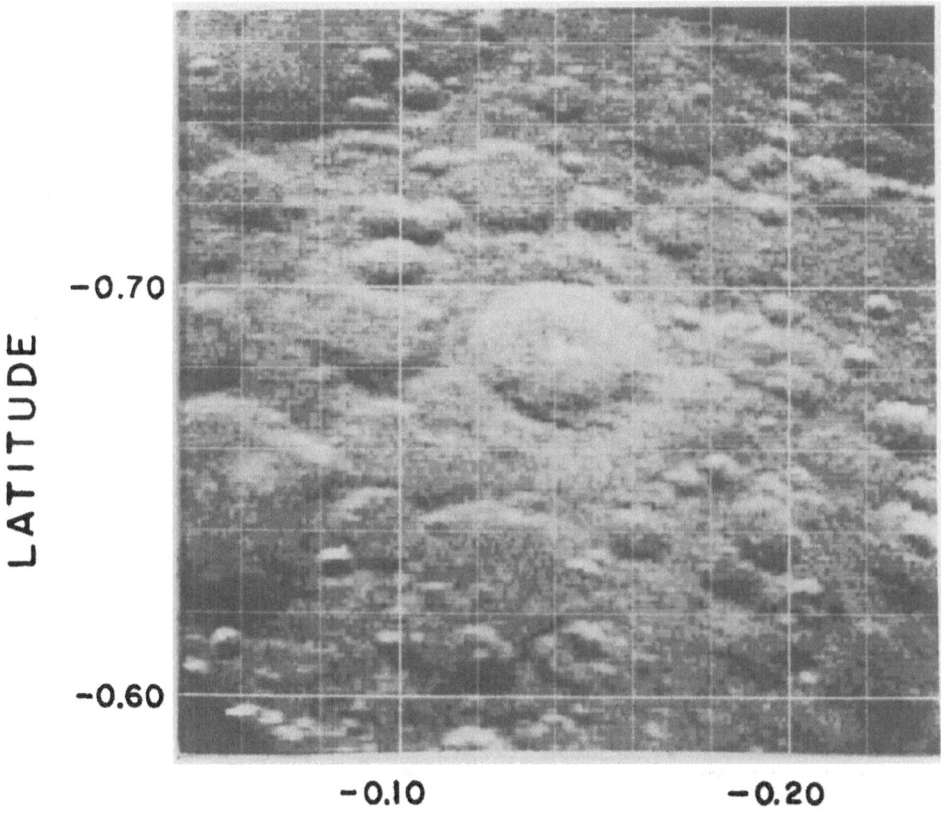

Fig. 7.1. Radar map of the lunar crater Tycho (see also Figures 1.30 to 1.32), reconstructed from observations at 3.8 cm wavelength in the polarized mode (after G. H. Pettengill and T. W. Thompson, *Icarus* **8**, 457, 1958). Surface resolution of this map is close to 1 km.

'radar maps' of the lunar surface attaining ground resolution comparable with the best that Earth-bound telescopic astronomy can offer at a distance (i.e., about 1 km on the lunar surface). This truly monumental work is only now being published (cf. Pettengill *et al.*, 1974a, b; Thompson, 1974) and one of their radar maps of the vicinity of the crater Tycho is reproduced on the accompanying Figure 7.1.

In comparing this map with other records of the same region obtained in other frequency domains, we should keep in mind that, at optical frequencies (i.e., in scattered sunlight), light and dark portions such as seen, for instance, on Figures 1.30–1.32 represent regions which are illuminated or in shadow; and on records obtained in the near IR (cf. Figure 7.2) they represent regions which are warm or cool. In radar illumination differences in intensity represent, on the other hand, *variations in reflectivity*. All portions of the lunar surface are 'illuminated' by the radar transmitter (thus approximating 'full-moon' conditions); but the contrast between bright and dark regions shown on radar pictures far exceeds that shown on full-moon photographs. On the latter, as we know, albedo differences between the brightest and darkest portions of the surface do not exceed a factor of three; but in the radar case, enhancement in reflectivity above the mean as large as by a factor of 10 have been observed.

How to account for such difference? There is little room for doubt that *the principal cause of differences in radar reflectivity is caused by different degree of surface roughness*. This is well borne out by the fact that not only enhanced reflections have

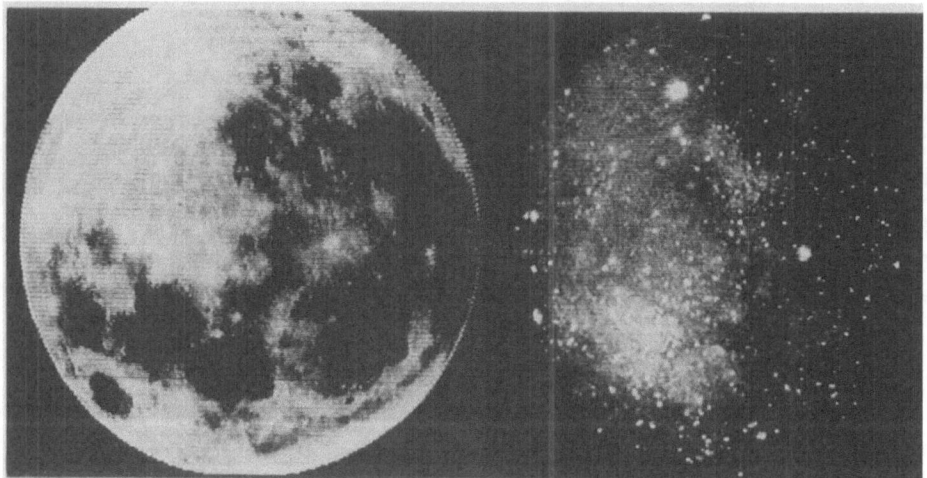

Fig. 7.2.   A composite image of full Moon, as reconstructed from photoelectric scans of the lunar surface in visible light with the 60-in. reflector of Mt. Wilson Observatory by Shorthill and Saari (left); while the image on the right shows a similar composite image of totally eclipsed Moon, based on infrared (10–12 μ) scans of the lunar surface during the eclipse of 19 December 1964 with the 74-in. reflector of the Kottamia Observatory in Egypt (cf. J. Saari and R. W. Shorthill, *Nature* **205**, 964, 1964). Each spot of the image on the right corresponds to a region of enhanced thermal emission, indicative of less rapid rate of cooling (the three most conspicuous coincide in position with the craters Tycho (top), Copernicus (right) and Langrenus near the left limb).

been obtained from craters whose floors have been shown by the Orbiter or Apollo photographs to be rough (in the case of Tycho, see Figure 1.32; but also by the discovery that the brightest parts of a crater on radar pictures are inward-sloping walls; and how rough or broken these can be was amply disclosed by such Orbiter photographs as are reproduced on Figures 1.31 and 1.32.

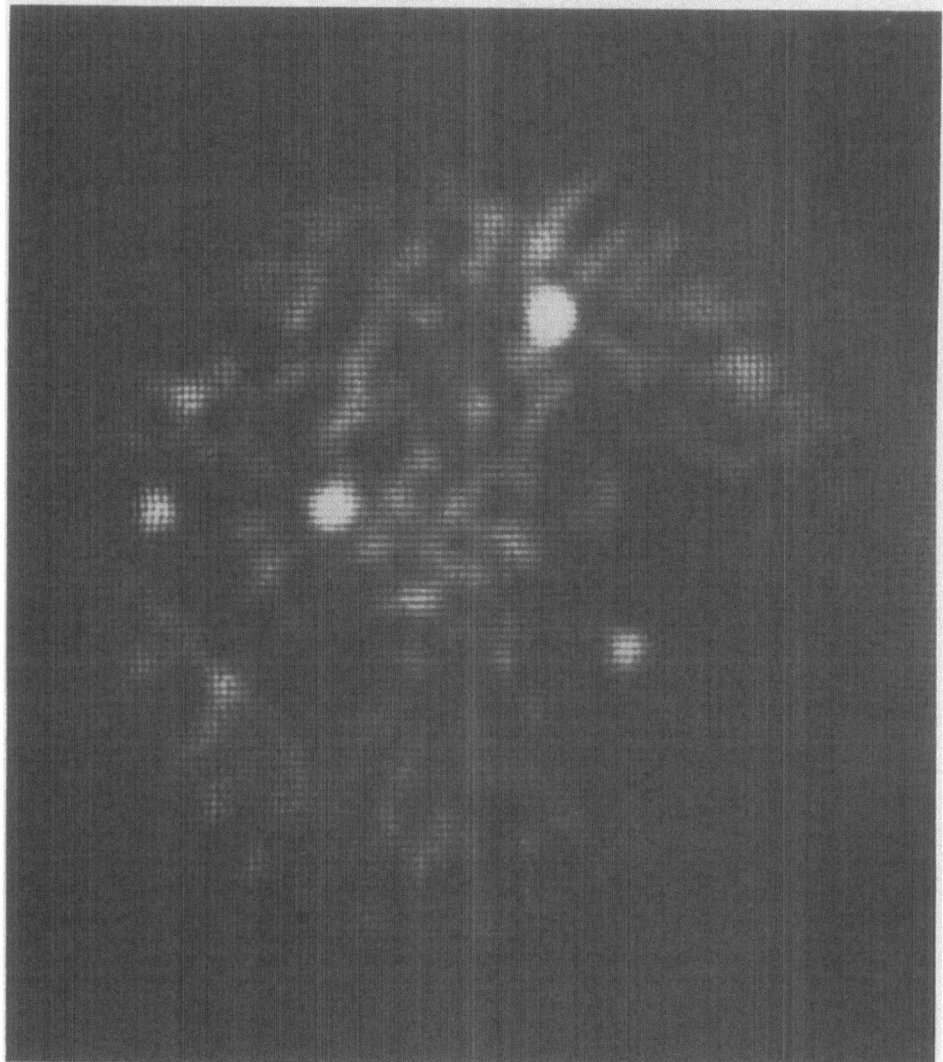

Fig. 7.3. The Moon as seen by radar in the depolarized component of the returning signals at 75-cm wavelength (originating at greater sub-surface depth than the returns on which Figure 7.1 was based). The four brightest spots of the record represent the craters Tycho (top), Copernicus (right), Theophilus (at the vertex of the triangle); with Langrenus to the left of it near the limb (after J. E. B. Ponsonby, I. Morrison, A. R. Birks and F. K. Landon, *Moon* 5, 286, 1972). The angular resolution of this radar map is 1¼′, corresponding to a linear resolution of close to 100 km on the surface. Therefore, the above-mentioned craters can appear on the map only as single picture-points.

Quite recently, aperture-synthesis radar method has been employed by Ponsonby and his colleagues at Jodrell Bank for mapping of the Moon at 75 cm and 185 cm wavelengths in depolarized light (cf. Ponsonby *et al.*, 1972) taking advantage of the libration of its surface. The results of their work are shown on the accompanying Figures 7.3 and 7.4, which disclose a number of interesting features. No significant

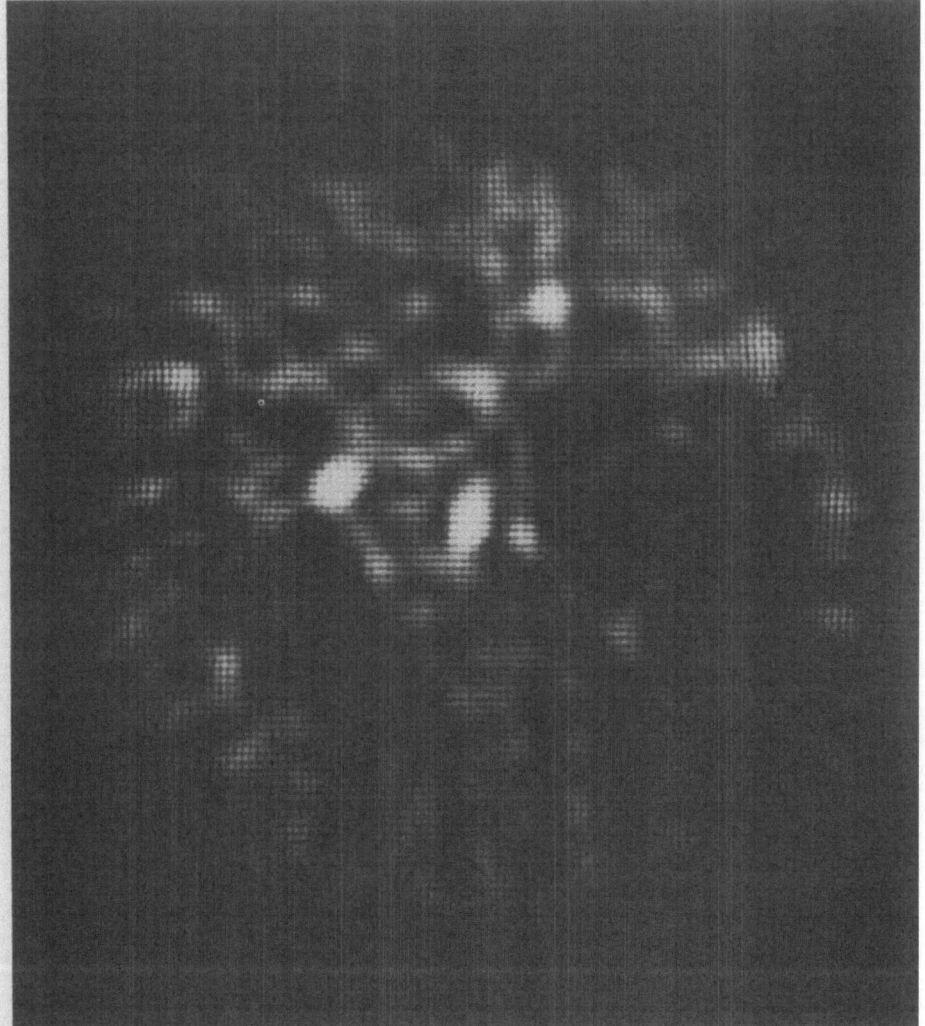

Fig. 7.4. The Moon as seen by radar in the depolarized component of the returning signals at 185 cm wavelength, originating still much deeper below the surface. The resolution of this map should be comparable with that shown on Figure 7.3; but as a result of the increasing depth of penetration, the record becomes quite different: at 185 cm wavelength the strongest return comes from Theophilus – much stronger than Tycho, let alone Copernicus or Langrenus. The new bright source (barely indicated on Figure 7.3) to the right of Theophilus is probably Ptolemaeus with Alphonsus.
(After Ponsonby *et al.*, 1972).

returns have been received from the maria, but certain isolated features – such as the craters Copernicus, Theophilus, or Tycho – appear to be particularly prominent; with Theophilus appearing brightest at the 185 cm wavelength, and Tycho at 75 cm – a fact which suggests that the surface structure of Theophilus (cf. Figures 8.1 and 8.5) is rougher on a 2 m-scale than is the case with Tycho, and (or) extends deeper into the ground.

This is, in brief, a preview of the structure of the lunar surface as it emerged from Earth-based observations made in the light of different wavelengths. The hard-landing U.S. Rangers 7–9 of 1964–1965 were the first spacecraft to provide – by means of their television cameras – the first glimpses of the lunar landscape on a 1-m scale (exceeding the resolution attainable from the Earth with our best telescopes by a factor of the order of 1000). A glance at photographs of the immediate neighbourhood of their respective landing places (cf. Figure 1.8) confirmed the anticipated smoothness of lunar ground, at least in the maria.

A more dramatic contribution to our knowledge of its structure has been provided by many soft-landing spacecraft since 1966. The priority in the distinguished series belongs to the records secured by Luna 9 (Figure 7.5) and Surveyor 1 (Figures 7.6 and 7.7), followed by subsequent Lunas (Figure 7.8) and Apollos (Figure 7.9). Of the five successful Surveyor missions, only the last (Surveyor 7) landed in a typically con-

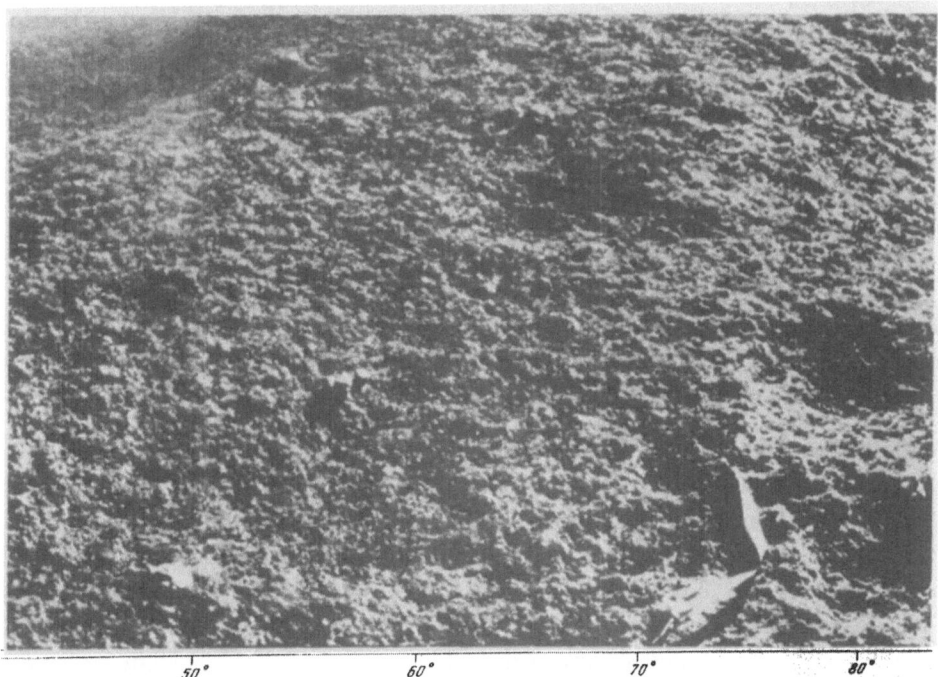

Fig. 7.5. A part of the horizon of the lunar surface televised from Luna 9 on 4 February 1966. The smallest details seen in close proximity of the spacecraft are only a few millimeters in size. (U.S.S.R. Acad. Sci. photograph).

Fig. 7.6.   Sunset over Oceanus Procellarum in the neighbourhood of the crater Flamsteed, televised by
Surveyor 1 on 13 June 1966 (NASA-JPL photograph).

tinental area (in the proximity of the crater Tycho); but a horizontal panorama seen by
its television cameras (see Figure 1.18) was not so different from that encountered in
the marial regions; though the ground in the immediate neighbourhood of the spacecraft appears to be rougher. Perhaps nothing can bring home the desolate nature of
lunar mare ground than a glance at the Apollo 11 photograph as shown on Figure
7.9, and featuring a view of the lunar landscape in the proximity of the landing place.

When the first soft-landers turned their TV-cameras to the immediate proximity
of the spacecraft – where they attained ground resolution of the order of 1 mm –
they discovered the scale-length at which the structure of the lunar surface changes
from smooth to rough, as anticipated by the photometric evidence. They found that,

Fig. 7.7.   A close-up view of the lunar surface at the time of sunset, televised by Surveyor 1 on 13 June 1966. Note the long shadow cast by a small boulder in the immediate proximity of the spacecraft; smallest details discernible on the photograph are only millimetres in size (NASA-JPL photograph).

commencing with the mm-scale, the lunar soil consists of granular material ranging considerably in size, and so loosely packed that the footpads of spacecraft (Figure 1.22) or of men walking over it (Figures 2.6 to 2.10) sunk into it to a depth of several centimetres. The details of such imprints disclose that lunar surface consists of material possessing a definite amount of cohesion, and not of loose dust.

The data provided by all soft-landers since 1966 – both manned and unmanned – lead to a conclusion that the bearing strength and density of the topmost lunar material changes rapidly with the depth. Thus for the upper few millimetres of the lunar surface the bulk density of (uncompressed) soil is between 0.7–1.2 g cm^{-3}; and the static bearing strength less than 10^4 dyn cm^{-2}. At a depth of about 2 cm, this bearing

Fig. 7.8. Rock-strewn surface of Mare Imbrium, televised by the first Russian Lunokhod (Luna 17) in 1971 (U.S.S.R. Acad. Sci. photograph).

Fig. 7.9. A view of the lunar landscape of Mare Tranquillitatis, photographed through the window of the Apollo 11 Lunar Excursion Module (the 'Eagle') in July 1969 (NASA-MSC photograph).

strength increases to 2×10^5 dyn; and further to 6×10^5 dyn cm^{-2} at some 10 cm below the surface, where the bulk density has risen to about 1.6 g cm^{-3}; and to 1.7 g cm^{-3} at a depth of half a metre (cf. Houston and Mitchell, 1974). Since, moreover, we know that the surface material are essentially silicates (of solid grain density close to 3 g cm^{-3}), the above-mentioned bulk densities at sub-surface depth between a few mm and a few cm would correspond to porosity factors diminishing from 0.8 to 0.5 over this depth. This is in satisfactory agreement with the expectations based on the observed dielectric properties of this material, as well as with the rapidity of the temperature changes – both diurnal as well as during eclipses – inferred previously from Earth-based measurements (cf. again Chapter 20 of the *Moon II*) and confirmed since by lunar spacecraft *in situ*. No man as yet witnessed an eclipse of the Sun on the Moon – but two unmanned spacecraft did so (Surveyor 3 on 1967 April 24, and Surveyor 5 on October 18 of the same year); and their data closely confirmed the surface temperature variations previously inferred from the Earth.

What may be the cause of this loosely-packed granular structure of the lunar surface, anticipated from the Earth-based photometry, and confirmed subsequently by spacecraft? Without wishing to discount other possible causes of more local nature, one obvious process which may have produced it – and which must be demonstrably operative on the Moon – is the constant downpour of micrometeorites swept up by the lunar surface on its perpetual journey through interplanetary space.

As is well known, since time immemorial the lunar surface has been exposed to continuous bombardment by all ingredients of the interplanetary space, ranging in mass from billions of tons for comets or asteroids down to the finest micro-meteoritic dust which has so far been tracked to particles of masses of the order of 10^{-17}–10^{-16} g*. The physical characteristics of these particles depend, in turn, largely on their mass. When the latter is measured in kilograms or more, the object is invariably a solid piece of stone (largely silicates), or iron, or a mixture of both. Between masses of approximately 10^2 and 10^{-8}, the particles can be broadly described as meteors, and derive largely from the icy nuclei of comets. On Earth (and, to a lesser extent, on Mars) meteor particles of this class spend themselves in the upper atmosphere and never reach the surface; so that their properties can be studied only by optical or radar methods. The population below this mass limit is composed of small particles of the dimensions measured in microns, which can be decelerated by collisions with air molecules without destruction. These are the true micrometeorites; and it is possible to collect them with the aid of high-altitude rockets, or retrieve them as they float down for weeks through the atmospheric molecular sieve to the surface of the Earth.

On the Earth (or any other planet surrounded by an atmosphere) only those particles can come into contact with the actual surface which can withstand atmospheric deceleration without total dispersal of their mass; and this can happen if the deceleration is either very small (i.e., large mass) making the time of flight short, or again

---

* Particles smaller still would be expelled from the solar system by solar radiation pressure and the Poynting-Robertson effect.

if the particles are so small that they can be decelerated already in the upper atmosphere without becoming too hot for vaporization. On the Earth, the prevailing air density tends to segregate such particles in two distinct classes: meteorites weighing 10^3 g and more, and micrometeorites of 10^{-7} g or less; particles of the intermediate mass range being effectively excluded from any contact with the terrestrial surface by the shielding effect of our air cushion. On the planet Mars, whose atmosphere is less dense than our own, the range of the masses which are thus forbidden access to the surface is smaller than 10^{10}, but still very large.

On the Moon, unprotected from the celestial intruders by any atmosphere to speak of (cf. Chapter 9), the solid surface is directly exposed to a continued infall of solid particles of all sizes and masses which the Moon intercepts on its perpetual journey through space. The influx of objects intercepted per unit area and time at the mean distance of the Earth from the Sun is now fairly well established for the entire particle population ranging from 10^{17} to 10^{-17} g in mass.

A most recent re-discussion of the underlying observational evidence has been undertaken by Soberman (1971) and Millman (1973); and the data collected by them are unlikely to undergo any important change in the future. In the preceding chapter of this book we were concerned with the scars which the impacts of interplanetary debris with masses m in the upper part of the observed frequency distribution (for $m > 1$ g) can produce on the lunar surface in the form of 'craters'. With impact craters of large and medium size range we already concerned ourselves in the preceding chapter. In what follows we shall confine our attention to 'micro-craters' on the lunar surface, caused by impacts of particles with masses $m \ll 1$ g – impacts which are perhaps less conspicuous, but very much more numerous; and, for this reason, of greater importance for the photometric properties of the lunar surface, if not for its topography.

What kind of 'etching' are such micrometeorites likely to produce on the lunar surface in the course of time? Two examples of micro-craters produced by such impacts are shown on the accompanying Figure 7.10. Their pits are only a few microns in diameter – and differ, therefore, by some 11 orders of magnitude in size from the largest lunar impact formations like Mare Imbrium, Serenitatis, or Orientale. The masses of the impinging particles (for relative impact velocity of 3 km s^{-1}) can be estimated to 10^{-14} g; and their size, to 0.2 μ. Nevertheless, even particles so tiny – swept up by the Moon from space in great numbers – can immortalize themselves by leaving behind microscopic craters whose characteristics can be studied with ease, and which – as we already heard – possess better chances of survival than the Egyptian pyramids on the Earth.

If we assume (to fix our ideas) that such particles impinge on the lunar surface with a velocity of (say) 10 km s^{-1}, the mass of particles capable of producing on impact micro-craters 1 or 10 mm in size turned out to be 3×10^{-9} and 7×10^{-6} g, respectively – corresponding to radio-meteors of essentially cometary origin. Moreover, the corresponding accretion times should be about 1 per mm^2 of the lunar surface every 300–400 yr, and 1 per cm^2 per 2–3000000 yr, respectively. This should be the time-scale on which the micro-relief of the lunar surface gets formed, destroyed, and re-created by external impacts of micro-meteorites raining down incessantly on the ground – in much the same way as raindrops on Earth may checquer the dry dusty surface with their pockmarks.

If this explanation of the roughness of the lunar surface on the mm-scale were correct, we should expect a rather sudden onset of roughness below 1 cm scale-length – corresponding to cratering by particles with masses close to 10^{-6} g. The mass-

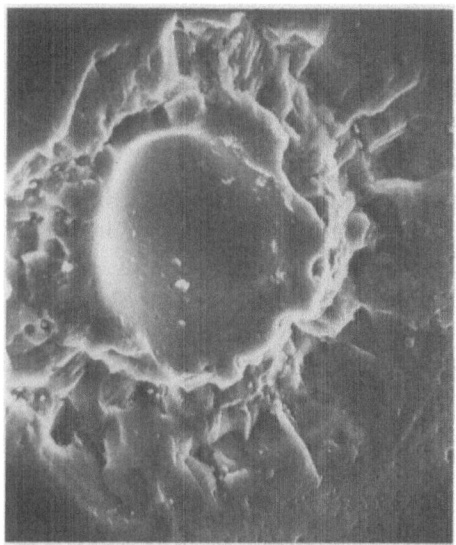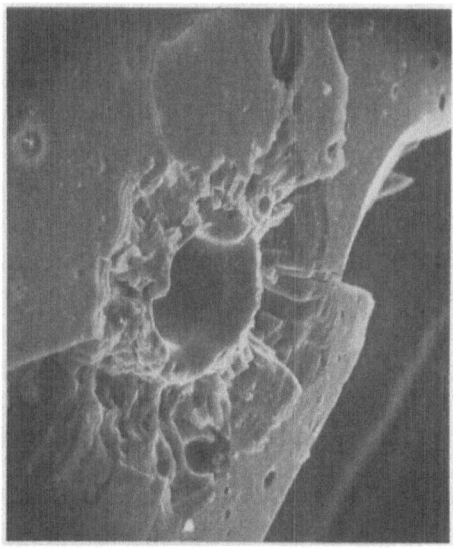

Fig. 7.10. Examples of impact-produced micro-craters in lunar rocks, brought back from the Moon by the Apollo 11 and 12 missions. The size of each field is approximately 10 microns across; and the central pits (a few microns in diameter) show clear evidence of melting by heat derived from the kinetic energy of impact. Photographs taken with a magnification of 2250×, and reproduced by courtesy of J. L. Carter of the University of Texas.

frequency distribution of interplanetary debris does show, in fact, a rather abrupt change in slope near this value (cf., e.g., McCracken and Dubin, 1964); but whether or not this represents more than a mere coincidence, only the future can tell. However, the fact that the salient features of lunar photometry (i.e., its phase-law and absence of limb-darkening) in the visible part of the spectrum are so different from those exhibited – for instance – by the planet Mars, can be attributed to the circum-stance that particles primarily responsible for surface micro-structure – in the 10^{-6}–10^{-9} g mass range – are (like on Earth) already destroyed by a passage through the Martian atmosphere, and thus filtered out from what eventually impinges on the surface of this planet.

In any comparison of the relative merits of micrometeoritic cratering and any other internal process which can be invoked to account for the vesicular or honeycomb structure of the lunar surface characterized by the requisite degree of back-scattering, a powerful argument for an external influence at work is provided by the photometric homogeneity of the entire lunar surface, demonstrating that micro-relief required to explain the observed light changes overlies all types of the lunar ground – in the maria, continents or bright rays – whatever their location or albedo. What else but an external influence could impress more readily the same uniform type of micro-relief all over the Moon?

The data bearing on the rate of meteoritic impact on the Moon lend themselves also for estimates of the total amount of meteoritic material (by integration of the observed influx as a function of the mass) which must have been so deposited on the lunar surface over long intervals of time. Öpik (1960) estimated the infall of cosmic debris per year at 10^{-8} g cm^{-2} – which, if extrapolated linearly over the entire lunar past of 4×10^9 yr, would lead to a total accumulation of some 40–50 g of material, per cm^2, corresponding (for a bulk density of rather less than one gram per cc) to a layer about one foot thick.

According to the latest estimates by Gault (1974), the topmost layer of half-mm depth gets turned over by micrometeoritic action, on the average, once in 10^4 yr; while 10^6 yr should be sufficient for a turnover down to a depth of 3 mm; 10^7 yr to 1 cm, and 10^8 yr to 10–100 cms. Only the topmost 0.5–1.0 mm veneer of a layer is subjected to intense churning and mixing at any one time; but this is also the layer responsible for most part of the optical properties (albedo, polarization) of the lunar surface at sub-micron wavelengths. It may (to a lesser extent) contribute to the observed thermal properties of topmost lunar crust; but assuredly not for the low values of its dielectric constant down to the level of penetration of long-wave radar pulses.

It goes without saying that this continuous action of micrometeoritic infall is occasionally interrupted by a much more intensive downpour of lunar material ejected by crater-forming impacts. The ejection of material by impacts which gave rise to such craters as Copernicus, Theophilus, or Tycho would have been sufficient (cf. Kopal, 1966) to cover the whole Moon with dust and coarser debris down to a depth of the order of 1 m; and such impacts as seem to have produced large circular maria – like Mare Imbrium or Serenitatis – could have (and apparently did!) bury the pre-existing landscape with their secondary ejecta down to a depth of many metres. But such impacts represent discrete and isolated events – while micrometeoritic infall constitute the principal source of erosion in between; and the effects of both must have been gradually diminishing in the course of time.

2. Chemical Composition

So far in this chapter we have been concerned with the physical and mechanical structure of the lunar surface on small scale. An investigation of the *chemical composition* of this crust was impossible before the advent of lunar spacecraft – beyond certain general surmises based on the known mean density of the lunar globe; and these we detailed in Chapter 11 of the earlier editions of this book.

The first lunar spacecraft which carried to the Moon automatic devices for analyzing the atomic composition of the topmost lunar crust was Surveyor 5 in September 1967, followed by Surveyor 6 (November 1967) and Surveyor 7 (January 1968); each landing in different parts of the lunar surface. The selenographic positions of their landing places have already been listed in Table 1-2) and the α-particle detector used for the analysis described in Chapter 1 (cf. also Figure 1.17).

The principal results of the analysis by Turkevich and his collaborators (cf. Turkevich *et al.*, 1968, 1969) are listed in the accompanying Table 7-1. These data repre-

TABLE 7-1

Atomic composition of the lunar surface at the Surveyor 5, 6 and 7 landing sites
(after Turkevich *et al.*, 1968, 1969)

Element	Surveyor 5	Surveyor 6	Surveyor 7
O	$61 \pm 2\%$	$57 \pm 5\%$	$58 \pm 5\%$
Mg	2.6 ± 1.5	3 ± 3	4 ± 3
Al	6.2 ± 0.2	6.5 ± 2	8 ± 3
Si	17 ± 2	22 ± 4	18 ± 4
Ca	5.9 ± 0.9	6 ± 2	6 ± 2
Ti	2.1 ± 0.8	5 ± 2	2 ± 1
Fe	3.7 ± 0.6		

sented the first direct information on the chemistry of the lunar surface; and although the method employed could disclose nothing on the molecular structure of lunar matter, by establishing a difference between the iron contents on the mare sites (Surveyors 5 and 6) and the continental area near the crater Tycho (Surveyor 7) it provided the first indication of the chemical *differentiation* of the lunar surface between different types of ground.

The real breakthrough in our knowledge of the chemistry of the lunar surface came with the advent of the re-entrant missions to the Moon (cf. Table 1-1) which returned to the Earth with their cargoes of lunar rock and debris collected with the human hand or automatic devices. The total rock payloads returned by the individual spacecraft are listed in the accompanying Table 7-2; and this material – amounting now

TABLE 7-2

Lunar rock payloads returned to the Earth by re-entrant spacecraft between 1969–1972

Mission	Payload (in kg)
Apollo 11	22
Apollo 12	34
Luna 16	0.101
Apollo 14	43
Apollo 15	78
Apollo 16	96
Luna 20	0.050
Apollo 17	113

to over 382 kg – has done more than anything else to open up for us the lunar past.

The literature which has resulted from a study of these 382 kg of rocks – probably the most thoroughly-studied lot in the entire field of cosmochemistry – since 1969 is already now truly enormous; and nothing but a brief survey of the salient facts which have emerged from it can be offered in this place. Data on the elemental composition of the lunar rocks and soils brought from different localities for the 10 most abundant elements of the lunar crust have been collected in Table 7-3 and compared with the corresponding composition of the terrestrial crust (last column); while Table 7-4 contains a similar compilation of their molecular structure. The abundances are presented in percentual proportion by weight; the balance being made up by trace elements (and compounds) occurring on the Moon in proportions of a few parts in a thousand or less. Most elements found on the Earth are also present on the Moon – including gold (in amounts of the order of a few parts per billion in weight); but their proportions indicate unmistakable differences from abundances found on the Earth's crust as to make it virtually certain – on cosmochemical grounds alone – that the material on the lunar surface never came from the terrestrial crust. And, last but not least, no trace of any organic compounds were found in the lunar samples that could indicate the presence of any life – recent or fossil – on the surface of our satellite.

An inspection of the data compiled in the preceding tables discloses that the most abundant element on the Moon – like on the Earth – happens to be oxygen, followed by silicon and aluminium. The fourth most abundant element – iron – is, however, present in the Earth and on the Moon in a markedly different proportion, and so are most other elements. Both iron and titanium contents appear to be distinctly enhanced in lunar crust; while the alkali metals are less abundant. Of other important elements, the same appears to be true of carbon, nitrogen, or hydrogen. An inspection of the molecular abundances in the Moon's crust discloses, furthermore, that all oxidized elements appear to be present only in their *lowest* stages of oxidation – a fact seemingly at variance with a very large proportion of oxygen listed in Table 7-3. It can, however, be understood in terms of the reducing action exerted on the lunar surface by the solar wind.

As is well known, the sunlit side of the Moon is continuously bombarded with corpuscular radiation evaporating from the Sun, whose predominant constituent is hydrogen. The protons – together with the accompanying electrons of a neutral plasma – travel from the Sun usually with velocities of the order of 300 km s^{-1}, and possess about 450 eV of kinetic energy. Before impact these protons are, however, neutralized by electrons evaporating from the lunar surface (cf. Chapter 9); so that what actually strikes lunar rocks is hot hydrogen gas, capable of reducing any known material to lower states of valence.

The reducing action of hot hydrogen is, in fact, so powerful that it could have virtually reduced the lunar surface to a metallic state. The reason why this has not happened is the low density of the solar wind. For, with the present flux of some 5 protons per cc travelling at 300 km s^{-1}, only some 75 million protons strike each

TABLE 7-3

Mean atomic composition of the lunar surface*

Element	Apollo 11	Apollo 12	Luna 16	Luna 17	Apollo 14	Apollo 15	Luna 20	Apollo 16	Apollo 17	Earth's crust
O	59.87	59.9	60.15	63.2	60.8	60.4	60.3	61.1	61.1	49.13
Si	16.31	16.0	15.97	15.7	17.4	17.3	16.0	16.3	16.3	26.00
Si	16.31	16.0	15.97	15.7	17.5	17.3	16.0	16.3	16.3	26.00
Al	6.30	6.3	6.95	5.7	7.7	6.5	9.7	11.6	10.1	7.45
Fe	5.12	5.4	5.39	4.7	3.1	4.5	2.1	1.6	1.8	4.20
Ca	4.92	4.1	4.99	4.4	4.3	4.4	5.9	6.1	6.1	3.25
Mg	4.57	6.8	4.99	6.3	5.4	5.9	5.2	3.0	4.0	2.35
Ti	2.19	0.9	0.98	1.8	0.5	0.5	0.15	0.15	0.15	0.60
Na	0.33	0.3	0.37	–	0.40	0.3	0.4	0.29	0.4	2.40

* After Turkevich (1972, 1973a, b).

TABLE 7-4

Mean molecular composition of the lunar surface*

Molecule	Apollo 11	Apollo 12	Luna 16	Luna 17	Apollo 14	Apollo 15	Luna 20	Apollo 16	Apollo 17	Earth's crust
SiO_2	40.70	44.95	43.8	48.0	46.07	42.40	45	47	48.5	48.5
FeO	17.42	20.53	19.35	10.5	21.19	6.40	7.5	8.6	10.5	10.5
CaO	10.52	10.94	10.4	10.7	10.21	18.20	13	12.1	10	10
TiO_2	11.00	3.32	4.9	2.1	2.13	0.38	0.8	1.5	2.1	2.1
Al_2O_3	9.43	9.19	13.7	17.1	8.95	20.20	23	21.2	16	16
MgO	7.34	9.83	7.05	8.7	9.51	12.00	8.5	9.9	7	7
Cr_2O_3	0.32	0.51	0.55	–	–	–	0.15	0.21	0.03	0.03
Na_2O	0.49	0.28	0.38	0.7	0.26	0.40	0.48	0.48	2.8	2.8
MnO	0.23	0.27	0.20	–	0.28	–	0.09	0.11	0.2	0.2
K_2O	0.18	0.058	0.15	0.5	0.034	0.52	0.20	0.15	1.2	1.2
P_2O_5	0.12	0.088	–	–	0.07	–	0.26	0.24	0.3	0.3

* After Gast (1972), Vinogradov (1972), and Schmitt and Laul (1973).

cm^2 of the lunar surface per second; which over the age of the Moon of 4.6×10^9 yr (cf. Chapter 8) could, on exidation, have covered the entire Moon with a layer of water some 80 cm deep.

On the other hand, what protects the Moon from total reduction is the photo-dissociation of water molecules by the UV light of the Sun. Under its action, water molecules dissociate into free hydrogen and hydroxyl radical OH. The former will quickly escape from the Moon's gravitational field (cf. Chapter 9), while the latter will oxidize rocks directly, or after decomposition into atomic oxygen and hydrogen. The lunar surface should, therefore, be at the mercy of a balance between the reduction tendency of the solar wind, and oxidation tendency of the solar UV light acting on water molecules.

How far did these processes advance in the past 4.6×10^9 yr? The salient fact is the low state of oxidation encountered in lunar rocks – and foreshadowed well before the advent of space age by a simple consideration of the colour of the lunar surface (or, rather, a lack of it) referred to already earlier in this chapter. We have known for a long time that the surface of the Moon appears to be almost uniformly grey – exhibiting (unlike Mars) no indication of any reddening which could be ascribed to hematite (Fe_2O_3); the ferrous oxide (FeO) figuring prominently in Table 7-4 is dark.

This fact discloses that, not only at the landing sites of different space missions, but all over the Moon *the effects of oxidation are conspicuous by their absence*; consequently, the Moon could never have had much more water on its surface than that imported by the solar wind. This fact is of considerable interest; for we know that the Earth exuded from its interior in the past an amount of water capable of covering the entire globe with a uniform ocean 1800 m in depth; and for the Moon (of 1.23% of the terrestrial mass) the same proportion of exuded water could have covered the surface (of 0.273 times the size of the Earth) with a uniform layer some 300 m deep. Yet this – if true – would have left the Moon's surface in a much more oxidized state than is consistent with its colour (or the oxidation of rock samples returned to the Earth from all localities). Therefore, the Moon could not have exuded from its interior anywhere nearly the same proportional amount of water as did our Earth; and the reason for this difference is no doubt the generally lower temperatures prevalent in the lunar interior, discussed already towards the end of Chapter 5; the Moon just has not developed enough internal heat to crack the bulk of its hydrates and expel the water outwards.

What kind of *minerals* do we find on the Moon to consist of the molecular units listed in Table 7-4, and what crystal structure do they exhibit? Before we attempt an answer, an analogy with the Earth may possibly be of some interest. On the Earth, all rocks occurring in the crust belong to one of the following three groups: (a) solidified sedimentary deposits from aqueous solutions; (b) metamorphic rocks – one-time sediments altered subsequently by pressure or temperature; and (c) igneous rocks which crystallized from cooling volcanic magmas. The latter can, in turn, be sub-divided into fast-cooling rocks (and exhibiting, therefore, fine structure of small crystals) – in contrast with slowly cooling rocks in which larger crystalline structures had a chance to develop. Basalts are typical examples of fast-cooling rocks, as granites are of slow-cooling crystalline structures.

Now, on the Moon, rocks of types (a) and (b) – sedimentary as well as metamorphic – are conspicuous by their absence; and in view of what we said in the preceding paragraphs about the lack of water in the lunar environment this must have looked even before 1969 like a foregone conclusion. All rocks brought back from the Moon are igneous – in the sense that they all solidified from molten magma at temperatures between 1100°–1200°C under highly reducing conditions (with partial pressure of free oxygen smaller than 10^{-13} atm). A discussion of the time and circumstances at which this occurred we shall defer till Chapters 8 and 10. For the present we wish to stress that crystalline rocks brought from the Moon consist almost entirely of minerals well known on the Earth – such as olivines, plagioclase, feldspar, ilmenite,

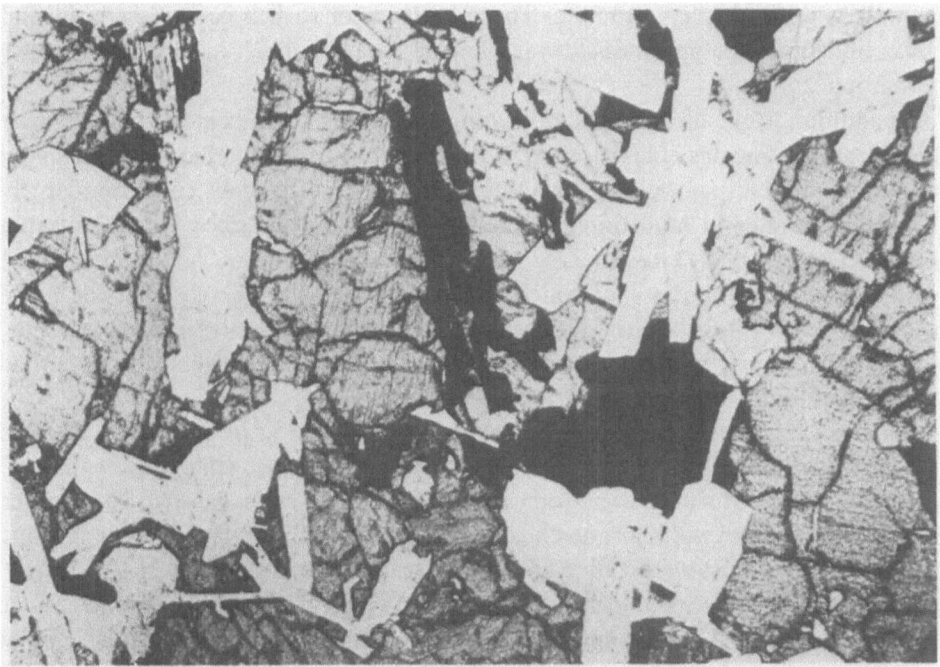

Fig. 7.11.   Apollo 11 sample No. 10044 of lunar crystalline rock under the microscope, showing the minerals pyroxene (gray), feldspar (white), ilmenite (black) and silica (crackled). Reproduced by courtesy of J. Zussman, Dept. of Geology, University of Manchester.

and many others (see Figure 7.11). Only three minerals not previously known from the Earth have been found in lunar samples so far – and none in any appreciable quantity.

The backbone of the dark crystalline material which covers the lunar mare basins can be loosely characterized as gabbroid basalts – akin to basaltic lavas on the Earth, but enriched with iron and titanium, and depleted in alkali elements. Moreover, the vapour phase associated with these lavas must have had higher CO/CO_2 and H_2/H_2O ratios than the terrestrial basalts. The phase-equilibrium arguments disclose that such basalts could not have been formed under pressures exceeding lunar hydro-static pressure (see Chapter 5) prevalent at a depth of some 400–500 km.

The bulk of the material (85–90% by weight) constituting the lunar continents appears to be *breccias* – polymixt conglomerates of pre-existing rocks, in which fragments of diverse origin which were welded together by multiple events subsequent to their solidification. The composition as well as structure of such breccias – exhibiting 100–1000 fold enhancement in the abundance of siderophile elements (Ir, Re, Rh, Au, etc., characteristic of the meteorites) with respect to lunar crystalline rocks (cf. Anders *et al.*, 1973), and strong evidence of shock metamorphism – leave no room for doubt as to the mode of their origin: they are the by-products of impact cratering of the lunar surface, which impressed on this surface its present gross topographic as well as mineralogical structure. The prevalence of such breccias in continental areas underlines the importance which impact cratering must have played in their formation.

In addition to the breccias, the remaining 10% of the rocks constituting the lunar continents (i.e., regions of higher reflectivity) appear to be feldspathic rocks, including a nearly pure feldspar called *anorthosite*. Anorthosites are rocks consisting largely of only one mineral (anorthite, of chemical structure described by the formula $CaAlSi_2O_8$), of density close to 2.8 g cm^{-2}. The latter is very much the same as that of the terrestrial granites – of which anorthosite may indeed represent the lunar equivalent. Anorthosites lack the iron or magnesium of basaltic rocks, having replaced them largely with aluminium; this is what makes them lighter in weight as well as colour.

The very existence of anorthosites – covering much the larger part of the Moon's surface (almost all of it on the far side) – implies some kind of chemical differentiation, in the course of which heavier elements – like iron – were separated from lighter ingredients. Moreover, lunar anorthositic rocks are mostly coarse-grained; and this means that they must have cooled off slowly – scarcely on the surface.

Lastly, among the smaller debris to be found in virtually all localities, we encounter samples enriched in potassium, rare earths, and phosphorus – ingredients which earned the material of this composition the playful name of KREEP (Hubbard and Gast, 1971). This 'cryptic' component constitutes an appreciable fraction (one-fifth to one-quarter) of lunar soil and is also represented in the breccias.

It should, however, be stressed that neither the mare basalts, nor the feldspar-rich rocks of the continental areas, possess structure and composition which would entitle us to conclude that the whole Moon may consist of such rocks. Both types would undergo phase changes to higher-density materials already at fairly shallow sub-surface depths, which would render the Moon consisting of them to possess too high a mean density. In order to maintain this density at 3.34 g cm^{-3}, compositional changes must be expected to occur with increasing depth.

In particular, the Moon as a whole must contain less metallic iron than the Earth, and its bulk composition must be quite different from that of the chondritic meteorites. The sources of the rocks we find on the surface must have been considerably enriched in refractory elements (Ba, Zr, Ti, rare earths) and depleted in chalcophile and siderophile elements as well as in the alkali metals. Such differences suggest that the primordial lunar material was subjected to elevated temperatures at the time of the crustal formation, or before its accretion.

This temperature could, however, not have been too high to melt these rocks completely. For apart from dynamical difficulties concerning the shape and moments of inertia in such a case, which we discussed already in Chapter 4 (and to which we shall return again in Chapter 10), total melting would also produce a complete mixing of molten material, resulting in the formation of rocks which are chemically homogeneous. Only partial melting can produce differentiation indicated by the available samples of lunar rocks. Fractional crystallization would also produce differentiation, but rapid cooling to be expected near the surface makes this process ineffective.

Turning back again to the atomic chemistry of the lunar surface we may note that, since Luna 10 and Apollo 15, rock samples returned to the Earth constitute no longer our sole source of information concerning the global chemical composition of the lunar crust. In particular, orbital science aboard the command module of Apollo 15 and 16 included two experiments of chemical significance: namely, X-ray fluorescence spectroscopy of the lunar surface (Adler *et al.*); and α- and γ-ray spectrometry to study the radioactivity of the Moon (Arnold *et al.*, 1972, 1973; Gorenstein *et al.*, 1974).

The X-ray fluorescence spectroscopy of the lunar surface has already been mentioned earlier in this chapter, in connection with attempts to observe such spectra from the distance of the Earth (cf. Juday, 1965). It constitutes essentially a passive experiment with an X-ray tube of cosmic dimensions, whose 'hot cathode' is the Sun, and the surface of the Moon playing the role of an anti-cathode whose chemical composition is under analysis. Because of the limited energy of incident 'solar wind', measurable X-ray spectra can be obtained only for lighter elements (up to, say; silicon); and heavier elements can be studied in this way only during brief intervals of more intensive solar activity.

An X-ray spectrometer to record the fluorescent spectra of the lunar surface in the domain of the X-rays excited by the Sun constituted a part of the equipment carried by the orbiting Luna 12 (cf. Mandelshtam *et al.*, 1968). Five years later, this work was resumed from the command module of the Apollo 15 and 16 missions. The angle of acceptance of the spectrometer, together with the altitude of the orbiting spacecraft, determined the resolution of the Apollo instrument to be no better than 10–20 km on the lunar surface; but with this resolution, the entire belt of lunar ground overflown by the respective spacecraft was available for analysis. Its most interesting result published so far has been the discovery of a close correlation of the ratios Al/Si and Mg/Si with the type of the ground overflown by the spacecraft: the abundance of Al being enhanced with respect to Mg over the continental (high-albedo) regions of the lunar surface, while over the mare-ground the converse proved to be the case.

In contrast with the X-ray analysis of the lunar surface where the energy of excitation came from outside, the α- and γ-ray spectroscopy of the lunar surface relies partly on indigenous energy sources: namely, γ-ray emission from uranium, thorium or potassium 40; or radon for α-particles. In addition, another class of discrete γ-ray emission can be expected as a result of the bombardment of certain elements (Al, Si, Fe) by galactic cosmic rays with energies of the order of 10^3–10^4 MeV (less energetic cosmic rays of solar origin are less important in this connection), giving rise to unstable nuclides on which more will be said later on.

The first γ-ray spectrometers to measure the emission by the lunar surface in this domain of the spectrum were aboard the early Ranger 3–5 missions, (cf. Arnold et al., 1962) but were unsuccessful because of the failures of some of these spacecraft. The first successful operation of this type of experiment we owe to the orbiting Russian Lunas 10 and 11 (cf. Vinogradov et al., 1968); but it was not till with Apollo 15 and 16 that this method had really arrived. Again, the spatial resolution was limited by the acceptance angle of X-ray optics and the altitude of orbiting command module to a circle of approximately 120 km, with the centre at the sub-spacecraft point.

The most striking result of this experiment has been the discovery of anomalous concentration of radioactivity in certain localized regions of the lunar surface – such as near Fra Mauro (the landing place of Apollo 14), or the region of Mare Imbrium – Oceanus Procellarum. Within the latter two distinct peaks in surface radioactivity were observed – one around the crater Aristarchus, the other in the south-eastern part of Mare Imbrium (see Figure 7.12). The level of radioactivity in these parts is by more than 30% higher than in the continental regions, where radioactivity remains generally low. The only notable exception to this appears to be the region around the crater Van de Graaf on the Moon's far side, where evidence of pronounced radioactivity is seen on Figure 7.12 (cf. Arnold et al., 1972).

It is, moreover, of interest to note that some regions of enhanced radioactivity (around Aristarchus, for instance) appear to coincide with where lunar transient luminous phenomena have been frequently reported in the optical part of the spectrum (cf. Chapter 22 of Moon II); but where this coincidence is merely accidental is for the future to decide.

In conclusion of our brief discussion of the physical structure and chemical composition of the lunar surface, mention must be made of another major contribution made by radio-chemistry to our knowledge of the processes of the lunar surface and its secular stability. The topmost layer of this surface is continuously exposed to irradiation (in daytime) not only by sunlight, but also by the entire corpuscular output of the Sun – ranging from particles (mainly protons) of the 'quiet-Sun' solar wind with energies less than 500 eV, to 'cosmic rays' of solar origin with energies in the 10–100 MeV range. What the former can do to the lunar surface we discussed already in the first part of this chapter; and now we wish to turn our attention to the more energetic – albeit scarcer – events occurring when a solar cosmic ray interacts with

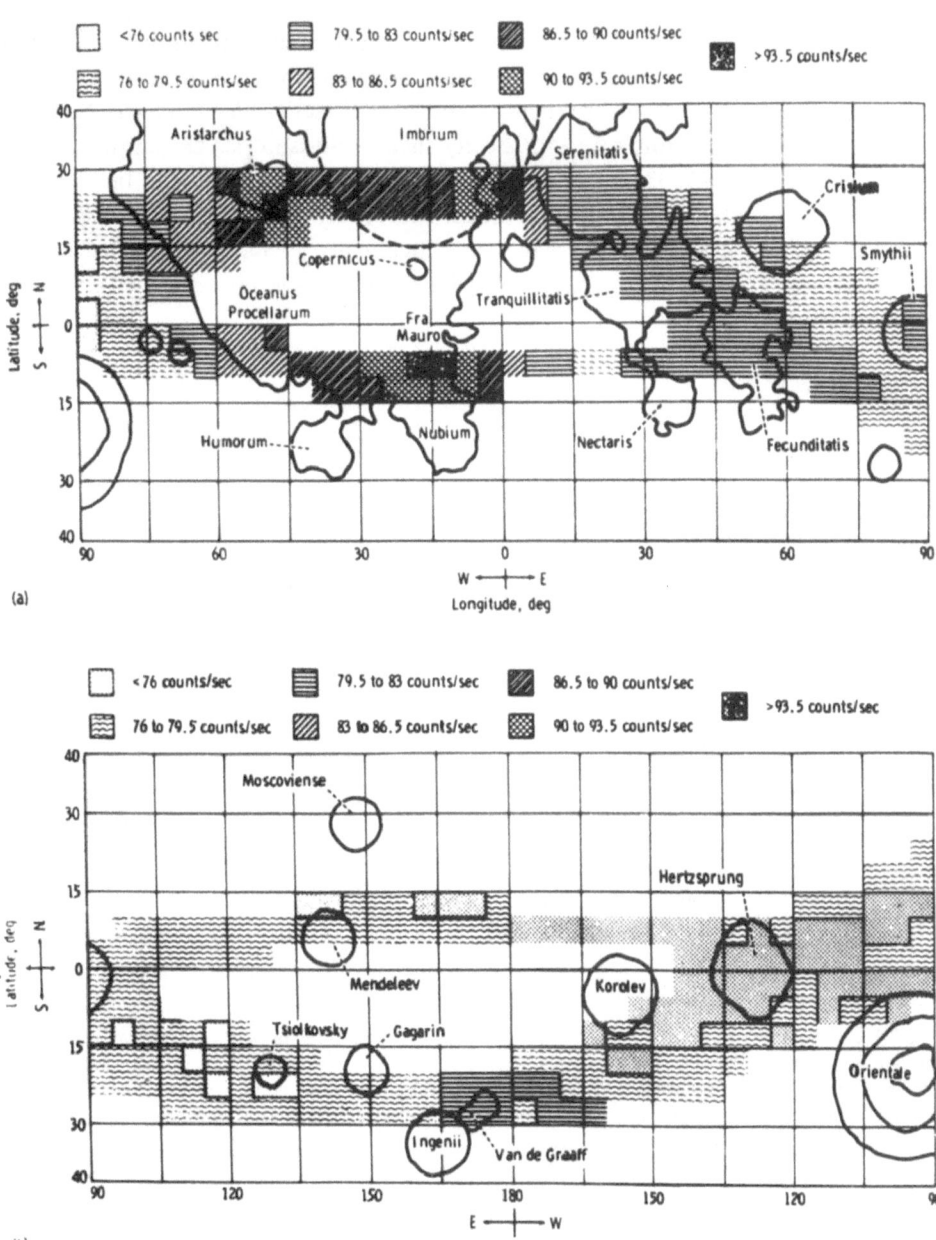

Fig. 7.12. Combined count rate results measuring the radioactivity of the lunar surface for Apollo 15 and 16 missions: (above) front side; (below) far side (after Arnold *et al.*, in *Apollo 16 Prelim. Sci. Rept.*, NASA SP-315, sec. 18, 1972).

the nucleus of certain light elements in the lunar crust. Such particles on impact lose their energy mainly by ionization. Sometimes, however, energetic head-on collisions give rise to nuclear reactions producing artificial radioactive elements with half-lives of the order of 10–100 million years. Many such radio-isotopes are known (for example, Na22, Al26, Mn53, Co56 or Ni59) and their concentration can be identified in lunar rocks by micro-chemical methods. Moreover, the impact of such particles can leave behind it latent tracks which are chemically 'etchable' in crystalline materials, caused by the solid-state damage due to ionization losses (see Figures 7.13 and 7.14).

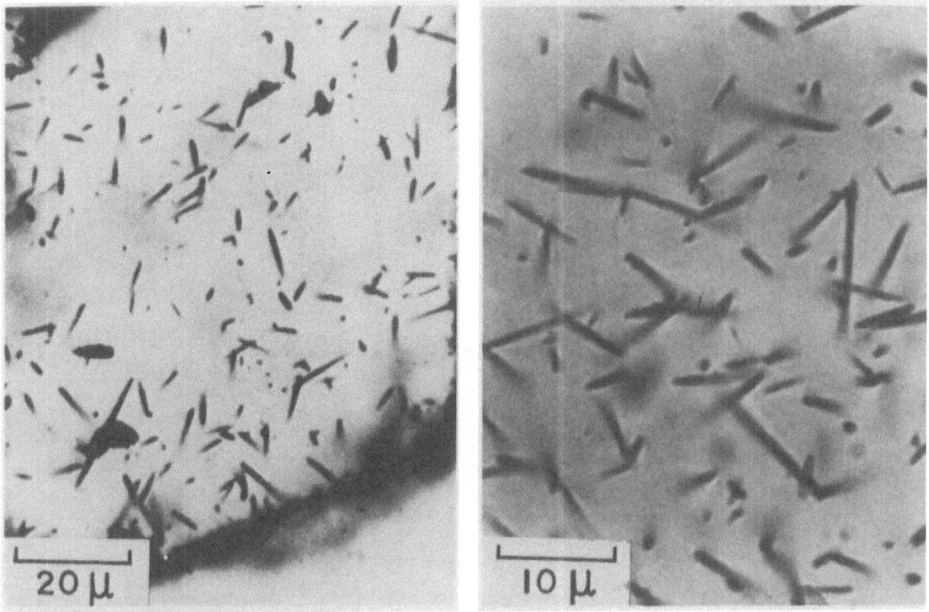

Fig. 7.13. Photomicrographs of fossil tracks produced by the passage of heavy cosmic-ray nuclei with charges $Z > 20$, as observed in feldspar crystals from Apollo 12 soil sample (left) and from the meteorite Moore County (right). The tracks are typically 5–10 μ long (and their holes have been decorated with silver to increase the optical contrast). Reproduced by courtesy of D. Lal of Tata Institute, Bombay.

Prior to 1969, such unstable nuclides, and particle tracks caused by cosmic-ray impacts, have been known to us only from the study of the meteorites (cf. Figure 7.14 right). Lunar rocks imported to the Earth since that time show likewise an extensive evidence of similar irradiation; and the concentration of artificial radioactive nuclides accumulated while the respective rock was exposed to cosmic rays or within their reach on (or near) the lunar surface can, therefore, be used to specify the 'exposure age' of the rock to cosmic rays. Moreover, an asymmetry in track distribution over the surface of such rocks can disclose the extent to which these rocks 'wiggled' in position or were overturned during this time.

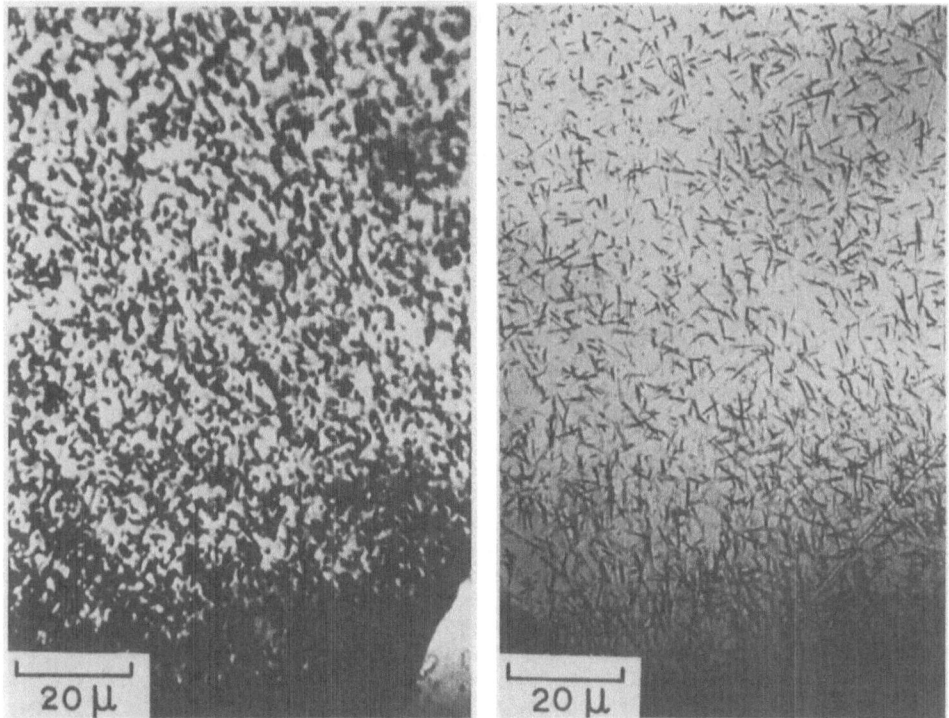

Fig. 7.14.   Solar cosmic-ray imprints in extra-terrestrial silicate grains. Photomicrographs of fossil tracks due to low-energy heavy nuclei accelerated during solar flares, as observed in the silicate grains from Luna 16 soil (left) and in the Kapoeta meteorite (right). Reproduced by courtesy of D. Lal, Tata Institute, Bombay.

The 'exposure ages' of lunar rocks brought from different localities (as distinct from the radioactive ages of solidification, to be discussed in the next chapter) turned out to run between 20–200 million years and more – which means that the respective rocks remained so long a time on the surface within reach of solar cosmic rays, until they were picked up by the hand of an astronaut. Cumulative effects of mechanical disturbances associated with impacts over longer intervals of time may bury rocks to sub-surface depths to which most cosmic rays can no longer penetrate – or again to exhume them from such depths; but a topmost layer reaching about 1 m below the lunar surface seems to be 'ploughed over' by mechanical or seismic erosion in a time-interval of the order of 100 million years – about the time separating us from the Cretaceous period on our Earth.

The extreme slowness of any changes on the Moon underlines the impression, gathered by a look at the lunar landscape as seen on Figures 7.6 to 7.9, of supreme desolation. The lunar plains are indeed more barren than any terrestrial deserts; and pitch-dark lunar caves are abodes of eternal silence. Above all, almost nothing ever happens on the Moon these days – at least before a host of spacecraft began to disturb its peace in the last 15 yrs – and changes prompted by processes discussed earlier in this chapter continue to grind the Moon's face exceedingly fine, but at an exceedingly slow rate. A spider's web stretched across a dim recess of any cavity would have a good chance to remain undisturbed for millions of years. Not a very exciting place for holiday-makers, perhaps; but for the scientist – what an Aladdin's cave full of heavenly wonders – both on the ground and above!

References

Adler, I. *et al.* (10 co-authors): 1972a, *Science* **175**, 436.
Adler, I. *et al.* (13 co-authors): 1972b, *Science* **177**, 256.
Adler, I. *et al.* (16 co-authors): 1973, *Moon* **7**, 487.
Anders, E., Ganapathy, R., Krähenbühl, U., and Morgan, J. W.: 1973, *Moon* **8**, 3.
Arnold, J. R., Metzger, A. E., Anderson, E. C., and Van Dilla, M. A.: 1962, *J. Geophys Res.* **67**, 4878.
Arnold, J. R., Metzger, A. E., Peterson, L. E., Reedy, R. C., and Trombka, J. I.: 1972, *Apollo 16 Prelim. Sci. Results*, NASA SP-315, 18-1.
Arnold, J. R., Metzger, A. E., Trombka, J. I., Peterson, L. E., and Reedy, R. C.: 1973, *Science* **179**, 800.
Gast, P. W.: 1972, *Moon* **5**, 121.
Gault, D. E.: 1974, in *Lunar Science* **V**, Pergamon Press, Inc., in press.
Gorenstein, P., Golub, L., and Bjorgholm, P.: 1974, *Moon* **9**, 129.
Houston, W. N. and Mitchell, J. K.: 1974, in *Lunar Science* **V**, Pergamon Press, in press.
Hubbard, N. J. and Gast, P. W.: 1971, in *Lunar Science* **II**, pp. 999–1020.
Juday, R. D.: 1965, *Trans. Amer. Geophys. Union* **46**, 142.
Kopal, Z.: 1969, in *Space Research* **IX**, North-Holland Publ. Co., Amsterdam, pp. 657–677.
Lynn, V. L., Sohigian, M. D., and Crocker, A. E.: 1964, *J. Geophys. Res.* **69**, 781.
Mandelshtam, S. L., Tindo, I. P., Cheremukhin, G. S., Sorokin, L. S., and Dimitriev, A. B.: 1968, *Kosm. Issled.* **6**, 119.
McCracken, C. W. and Dubin, M.: 1964, in J. W. Salisbury and P. E. Glaser (eds.), *The Lunar Surface Layer*, Acad. Press, New York and London, pp. 179–214.
Millman, P. M.: 1973, *Moon* **8**, 228.
Öpik, E. J.: 1960, *Monthly Notices Roy. Astron. Soc.* **120**, 404.
Pettengill, G. H. and Henry, J. C.: 1962, *J. Geophys. Res.* **67**, 4881.
Pettengill, G. H., Zisk, S. H., and Thompson, T. W.: 1974a, *Moon* **10**, 3.
Pettengill, G. H., Zisk, S. H., and Catuna, G. W.: 1974b, *Moon* **10**, 17.
Ponsonby, J. E. B., Morison, I., Birks, A. R., and Landon, F. K.: 1972, *Moon* **5**, 286.
Rossi, B.: 1973, in H. Bradt and R. Giacconi (eds.), 'X- and Gamma Ray Astronomy', *IAU Symp.* **55**, 1.
Schmitt, R. A. and Laul, J. C.: 1973, *Moon* **8**, 182.
Soberman, R. K.: 1971, *Rev. Geophys. Space Phys.* **9**, 239.
Thompson, T. W.: 1974, *Moon* **10**, 51.
Thompson, T. W. and Dyce, R. B.: 1966, *J. Geophys. Res.* **71**, 4843.
Turkevich, A. L.: 1973a, *Accounts of Chem. Res.* **6**, 81.
Turkevich, A. L.: 1973b, *Moon* **8**, 365.
Turkevich, A. L., Franzgrote, E. J., and Patterson, J. H.: 1967, *Science* **158**, 635.
Turkevich, A. L., Franzgrote, E. J., and Patterson, J. H.: 1968, *Science* **162**, 117.
Turkevich, A. L., Franzgrote, E. J., and Patterson, J. H.: 1969, *Science* **165**, 277.
Vinogradov, A. P.: 1968, in A. Dollfus (ed.), *Moon and Planets* I, 71; II, 77.
Vinogradov, A. P.: 1972, *Trans. Amer. Geophys. Union* **53**, 820.

CHAPTER 8

STRATIGRAPHY AND CHRONOLOGY
OF THE LUNAR SURFACE

In the preceding two chapters we discussed briefly the nature of the surface of the Moon as well as of the principal types of formations encountered on the lunar landscape. The aim of the present chapter will be to consider now the *collective* aspects of this latter evidence – the distribution of such formations over different types of lunar ground – in an attempt to sort out their stratigraphic sequence which could be provided with an absolute time calibration.

We stressed already in earlier parts of this book that the origin of all features visible on the surface of our satellite must go back to either the internal processes (for which this surface represents merely the outer boundary condition), or to external impacts of particulate matter of all sizes – from micrometeorites to comets or asteroids – bombarding it from outside (and for which the lunar surface constitutes an 'impact counter'). At the commencement of this chapter we wish to explore further the consequences of an assumption that the dominant process responsible for shaping up the macroscopic lunar relief has been the cratering by external impacts. As the Moon is virtually devoid of any atmosphere (cf. Chapter 9), and its surface is directly exposed to impacts of all particles which happen to be obstructed by the Moon in their heliocentric orbits, a bombardment of the lunar surface by meteorites and other bodies floating through space must demonstrably occur.

Moreover, the scars left by such impacts will fall no prey to erosion (other than seismic) which could gradually obliterate older features; pre-existing formations can be effectively destroyed only by fresh impacts. If so, however, the cumulative effects of such impacts – the fossil record of which has been preserved on the lunar surface with a permanence greatly transcending that of any geological markings on the Earth – should enable us to discern on the Moon a *stratigraphic time-sequence*, reflecting events of bygone days which could take us far into the past of the solar system – a record which we do not possess on the Earth because of a greater internal activity of our planet; and herein rests the scientific significance of such studies.

The adoption of such a point of view as a basis for statistical approach to our problem should not be construed to imply that we regard all formations seen on the Moon as being of external origin. Far from it; for internal causes must have been operative in parallel to produce many types of formations (domes, rilles, or wrinkle ridges, for example) which are manifestly impossible to account for by direct intervention of external agents. Nevertheless, we do believe that the largest and most conspicuous formations on the Moon are of external origin, and that many of them belong among the oldest landmarks on the surface of our satellite; the problem is to sort them out in a consecutive time sequence.

There are, indeed, three different and independent (though indirect) ways which can be invoked to this end: namely, (a) the principle of overlap; (b) the degree of ruggedness; (c) the ground reflectivity; and in what follows we shall discuss them in turn. Of the three, the 'principle of overlap' is perhaps the most obvious and dependable. If there are two craters that overlap each other – and the photographs reproduced in this article show many examples of such a situation – then the one with unbroken rim must be more recent than the one whose rim was damaged or entirely removed. In the case of two overlapping craters of comparable dimensions – such as the pair of Theophilus and Cyrillus on Figure 8.1, or the Vavilov brothers shown on Figure 2.18, for instance – an application of this principle requires the formation of Theophilus to have been posterior to that of Cyrillus. But the principle can be applied also to a situation in which a large crater contains smaller ones within its enclosure (for an illustrative example, cf. a photograph of the crater Clavius reproduced on Figure 8.2. No kind of process that raised the ramparts of Clavius would have left the smaller formations now seen within it undisturbed. Therefore, small craters on the floor of Clavius must represent creations subsequent to that of the large configuration. On this reasoning Clavius must be considerably older than all other craters which can at present be seen on its floor. The greater the number of small craters inside a large one, the greater should be the diparity between their ages.

Crater overlaps of great multiplicity can be found in certain part of the lunar surface; and in some places it is possible to arrange thus five or six craters in a time sequence. Another relative age criterion (supplementing overlap) is afforded by the existence of the streaks of relatively bright material (the 'rays') which are seen diverging in all directions from certain impact craters. As these rays must have originated by ejection at the same time as the parent crater, the latter must obviously be younger than any features overlaid by its rays. Such rays represent, indeed, a system of tentacles spreading widely over certain parts of the lunar surface and enabling us to extend our system of relative dating as far as they reach.

The main importance of the preceding age criteria lies in their application to the dating of the maria. If we accept the foregoing premises, there seems no escape from the conclusion that *the oldest parts of the visible lunar surface are those which are most rugged*, and contain the greatest number of craters or other types of mountains per per unit area. For irrespective of whether the rate of operation of both the external or internal processes of crater formation on the Moon has been uniform or diminishing with time, the oldest parts of its surface should obviously have accumulated the greatest number of scars. If this is indeed so, then the oldest regions of the Moon abound on the Moon's far side; and, on the visible hemisphere, those surrounding the Moon's south pole. Conversely, as the mean density of craters per unit area of the marial ground is much less than that encountered near the Moon's south pole, it follows that lunar maria should be younger than the continental regions; and some

Fig. 8.1.   Sunset over a group of lunar craters on the Eastern shores of Mare Nectaris, consisting of Theophilus, Cyrillus, and Catharina. The walls of the former two overlap in a way which demonstrates that Cyrillus must be the older than Theophilus. Photograph taken on 29 July 1964 with the 24-in. refractor of the Observatoire du Pic-du-Midi (Manchester Lunar Programme).

Fig. 8.2.   The lunar crater Clavius – the largest formation of its type visible from the Earth – photographed by the 200-in. Hale reflector of the Mt. Wilson and Palomar Observatories, Carnegie Institution
of Washington and California Institute of Technology.

of the great craters like Copernicus, – which spread the tentacles of their bright rays over large parts of the surrounding maria – must be still younger.

In point of fact, astronomical as well as physical and chemical reasons can be listed why the brighter any element of the surface appears, the more recent it is likely to be; for the passage of the time tends only to darken the exposed ground. This process lends itself, in turn, for interesting applications to a study of the stratigraphy of impact craters. As we mentioned already, many craters of this group are the foci of prominent ray patterns, but others are entirely unaccompanied by rays. Thus Eratosthenes is a good example of a crater that exhibits all the principal topographic features of Copernicus and is surrounded by a well-developed pattern of gouges, but· completely lacks rays. Moreover, where not overlapped by the Copernican ejecta, the walls as well as the floor of Eratosthenes exhibit relatively low reflectivity.

All gradation may be observed in the apparent brightness of the rims and associated rays among craters of impact type. Copernicus, Aristillus and Theophilus represent a sequence of craters accompanied by rays of diminishing reflectivity. The rays of Aristillus are plainly visible, but not as bright as those of Copernicus; while the rays of Theophilus are already very faint (cf. Figure 8.5), though its secondary impact craters are as widely distributed and as numerous as those of Copernicus. The reflectivity of the walls of Theophilus approaches that of Eratosthenes.

It is highly probable that this sequence reflects an increasing age of the respective formations. Wherever a Copernican-type crater without rays (or a crater with very faint rays) occurs in an area traversed by rays diverging from some other crater, the bright rays are in all cases superimposed on the darker crater or the fainter ray pattern; no single instance of a converse case is known. Some process, or combination of processes, must then evidently be at work on the lunar surface that causes the gradual fading of the rays and other surface elements of higher reflectivity. The infall of cosmic dust is indubitably one – and probably the most important one – in the course of time; but darkening of the material itself by radiation damage and mixing of the thin layer of ray material with underlying darker base by micrometeoritic bombardment may also contribute to the general process of fading.*

Prior to the advent of spacecraft which could return to us from the Moon with their cargo, any idea about the *absolute time-scale* of this tentative system of lunar startigraphy could have been obtained only by a comparison of the areal density of the impact structures on the Moon with the corresponding values established on the Earth. The method was inherently weak, because most part of the terrestrial evidence is severely short-lived (as erosion by air and water will obliterate completely most impact landmarks within less than one million years). In addition, a large fraction of the terrestrial surface has been protected from impacts by the oceans; so that the total numbers of the available sample are small. In order to apply them to the Moon,

---

* It may be noted, in this connection, that the albedo of interplanetary dust constituting the zodiacal cloud, which is constantly exposed to solar radiation damage at a much closer range, is only about 0.02 – i.e., more than three times smaller than the mean albedo of the Moon, and one-half of that of the darkest spots on the surface of our satellite.

an extrapolation by two or three orders of magnitude in time was necessary – without any assurance whether the flux of the impinging material remained constant, or diminished in the course of this time. Its constancy is far from likely, but was the best we could assume only a few years ago; and the results based on this assumption were sufficient to indicate that the origin of most part of lunar landscape must go back to the archaic times on the Earth (cf. Shoemaker and Hackman, 1962; Hartmann, 1970).

The real breakthrough – perhaps the most important single scientific contribution of re-entrant lunar spacecraft from 1969–1972 to solar-system studies – has been to bring back to the Earth extensive samples of lunar rocks, collected at different lo- calities of the lunar surface whose absolute ages could be ascertained by radiometric methods. As has been known from the beginning of this century, the absolute ages of solid rocks can be established from the progress of spontaneous disintegration of certain chains of radioactive elements that are present in such rocks in measurable amounts. A list of the principal types of 'nuclear clocks' used for this purpose is given in the accompanying Table 8-1, which contains the 'mother' and 'daughter' products

TABLE 8-1

Spontaneous nuclear disintegrations
used for the determination of age

Cycle	Half-life (in 10^9 yr)
U^{238}–Pb^{206}	4.51
U^{235}–Pb^{207}	0.713
Th^{232}–Pb^{208}	13.9
Rb^{87}–Sr^{87}	47.1
K^{40}–Ar^{40}	1.27

of the respective chains, together with the half-lives of distintegration (i.e., times in which the amount of the 'mother' substance will spontaneously disintegrate to one- half). These half-lives range from 713 million years for uranium-235 to 47 billion years for the rubidium-strontium β-decay. But all these clocks possess one feature in common: namely, they start marking time from the moment when the rock contain- ing them has solidified. For the dials of these 'radiometric clocks' (which indicate the proportion of the daughter product to the mother substance) are automatically set back to zero whenever the respective mineral has been remelted – just as (on a time- scale six orders of magnitude shorter) the well-known carbon-14 method of dating can tell the archaeologist the age of any organic tissue, measured from the time when the organism in question stopped its intake of cosmic-ray produced C-14 from the atmosphere by breathing at the time of its death. But, in comparison, the astronomer measuring the age of a lunar rock has one advantage; for while the measurements of time by the carbon-14 method is subject to the uncertainty in fluctuations of the cosmic-ray flux (at the source, or because of the fluctuations in the magnetic field of the Earth), the rate at which radioactive elements continue disintegrating on the Moon progresses at a constant rate, supremely oblivious of their surroundings.

When such radiometric methods were applied to the samples returned from the Moon by successive Apollo and Luna missions between 1969–1972, the *mean* ages of rocks brought back from different localities turned out to be as listed in the accompanying Table 8-2. The outstanding fact emerging from these data is the extremely high age of the lunar material in general – indeed, no rock has been found

TABLE 8-2

Mean ages of lunar rocks from different localities

Mission	Region	Age (in 10^9 yr)
Apollo 11	Mare Tranquillitatis	3.7
Luna 16	Mare Foecunditatis	3.4
Apollo 12	Oceanus Procellarum	3.6
Apollo 14	Fra Mauro	4.0
Apollo 15	Mare Imbrium	3.3
Apollo 16	Descartes	4.1
Luna 20	Mare Foecunditatis	3.9
Apollo 17	Taurus-Littrow	3.8

on the Moon in more than one locality whose radiometric age would be less than 3.2 billion years. How different from what we find on the Earth!

Before we attempt to absorb the full importance of this fact for lunar studies, we wish to caution the reader not to overestimate the significance of the differences in ages between different localities listed in Table 8-2. Each age given there represents merely the maximum of the age distribution of rocks imported from each respective locality; and the wings of the histograms of each distribution overlap to a considerable extent. Indeed, in view of the nature of the processes operative on the lunar surface, it would be unrealistic to expect to find only 'native' rocks in any exposed locality – contributions from many parts of the Moon should have been made to every element of the surface by impact transfer. In some parts of the Moon – such as the maria – specimens of local origin (mare basalts) may outnumber imported rocks in any sample selected at random. On the other hand, virtually all sites visited by man in recent years lie in the proximity of the Imbrian basin; and the impact which created this mare threw up so much material as to submerge all localities within its reach under a layer of ejecta many metres in depth; and samples brought back from such localities should be heavily intermingled with them.

Not only can local samples thus be contaminated by extraneous material from other localities (not to speak of meteoritic infall); but also individual rocks constituting such samples show often complicated time-history. The radio-chemistry of many rocks brought to light various disturbances of their isochrones – indicating that 'radioactive clocks' were tampered with by Nature more than once in the past.

The K-Ar clock can easily be reset by the diffusion of argon; for an exposure of lunar rocks to a temperature of 700 °C could lead to a total loss of radiogenic argon in a time-span of approximately one year; and within 10^4 yr at a temperature of 300 °C. The re-setting of the Rb-Sr clocks could again be caused by a homogenization of material under such conditions in approximately the same time; and it is probable that an appreciable fraction of rocks collected on the Moon could have experienced such disturbances.

But perhaps the most interesting feature of radiometric ages of lunar rocks has been the establishment of a systematic difference in age between the impact-generated breccias and mare basalts. Indeed, if we identify the time which elapsed since the mare-forming impacts of the principal lunar multi-ring basins with the age of the associated breccias, we find (cf. Schaefer and Husain, 1974) that these form a sequence listed in Table 8-3.

TABLE 8-3

Radiometric ages of
mare-forming events

Mare	Age (in 10^9 yr)
Nectaris	4.20 ± 0.05
Humorum	4.16 ± 0.04
Crisium	4.13 ± 0.05
Imbrium	4.00 ± 0.05
Orientale	3.85 ± 0.05

On the other hand, the radiometric ages of lunar mare basalts range only from 3.3 to 3.8 billion years – i.e., are substantially *smaller* than the age of the associated breccias – a fact indicating that *the flooding of* (some) *mare basins by basaltic magma must have taken place considerably later than the basin-forming impacts*. But even the initial formation of these basins does not seem to have occurred at approximately the same time, but to be spread over a few hundred million years; and a frequent occurrence of the 4.0 billion-year age among lunar rocks from different Apollo sites is probably due to a preponderance of the Imbrium ejecta on the topmost surface most likely to be sampled. In contrast, the Orientale event (the youngest of all) does not seem to have covered the same ground with anywhere nearly the same layer of ejecta.

The significance of these facts for our views on the evolution of our satellite will be discussed in the last chapter of this book. At present we wish merely to reiterate that the principal pattern of the lunar maria as we see them on the Moon's face today must have been formed between 3.3 and 4.2 billion years ago, as a result of an apparently unrelated series of distinct events which occurred within that time; and nothing of comparable importance is recorded in the lunar chronology to have occurred at a later date.

This does not mean, of course, that no macroscopic events happened on the Moon since that time; for our satellite has continued to act as a target of all external impacts – cometary as well as asteroidal or meteoritic – which collided with its surface in the past three billion years. We have reasons to believe that the frequency of such impacts greatly diminished since the preceding times (because the supply of stray bodies which could run into the Moon was gradually getting exhausted); and that nothing comparable with the 'saturation bombardment' which disfigured most part of the Moon's far side (or of the polar regions of its visible hemisphere) took place after the formation of the maria. However, we also possess now a reliable evidence that some of the more spectacular impact craters on the Moon – such as Copernicus,

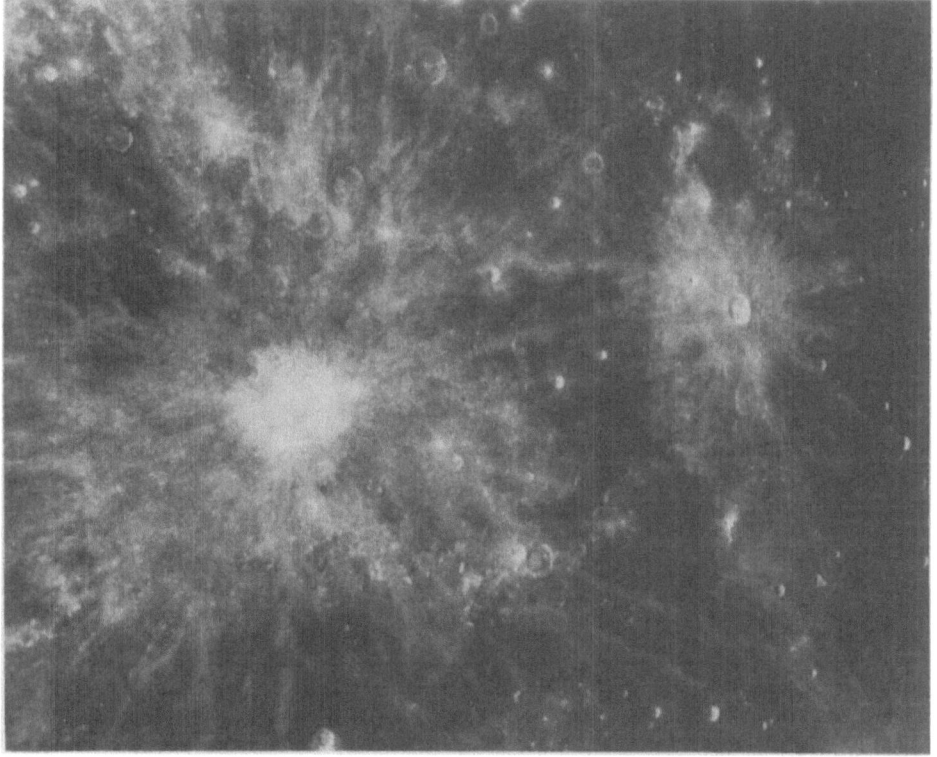

Fig. 8.3.   Bright ray systems surrounding the craters Copernicus (left) and Kepler (right), as photographed by the 24-in. refractor of the Observatoire du Pic-du-Midi (Manchester Lunar Programme).

Theophilus, or Tycho (see Figures 8.3 to 8.5) – were formed much later than the principal sculpture of their surrounding ground.

Earlier in this chapter we mentioned the advantages we can derive from the 'principle of overlap' for the relative dating of lunar features; and one of the best tools which lend themselves for this purpose are the splash phenomena (bright rays) which surround the regions of fresh impacts and spread their tentacles over sometimes considerable areas of adjacent lunar ground. The ray pattern of Copernicus (Figure 8.3) is a conspicuous example of such a phenomenon; and there are reasons to believe that Apollo 12 landed in the proximity of one such ray whose material found its way in our laboratories. A recent determination of its age (by the Ar^{39}/Ar^{40} method) at the hands of Eberhardt and his collaborators (Eberhardt *et al.*, 1973) established for this particular sample a value close to 800 million years – i.e., about a quarter of the

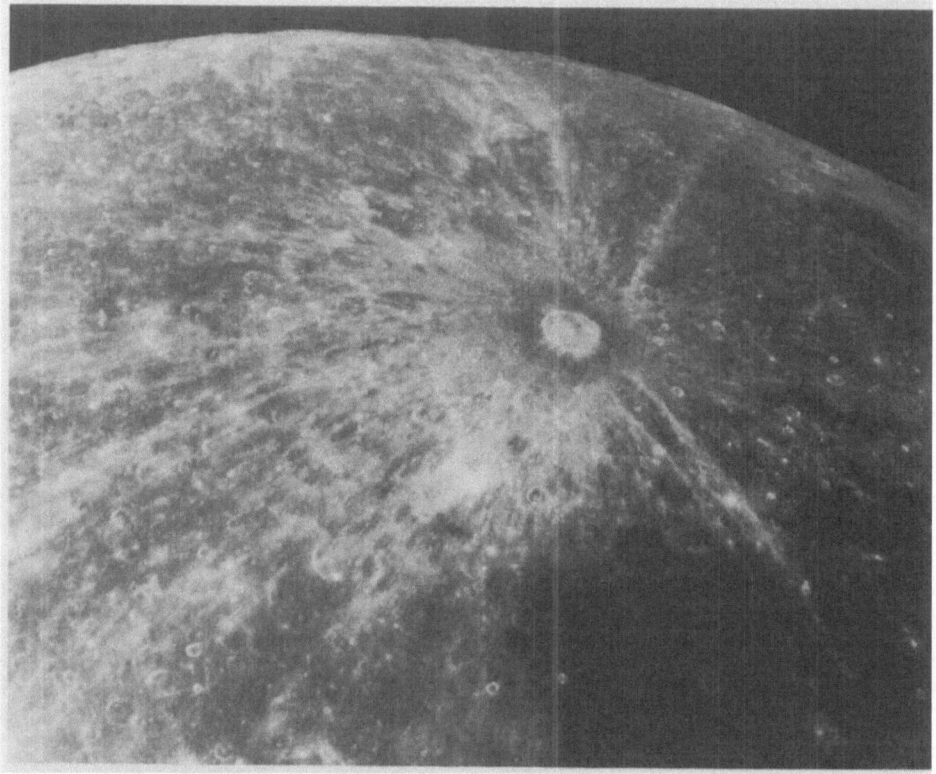

Fig. 8.4. A system of bright rays diverging from the crater Tycho near the Moon's south pole. Note the relatively dark annulus surrounding the ramparts of this crater (showing the extent of the 'ballistic shadow' for the ejecta), and also some rays which run tangentially to the walls. Photograph taken with the 24-in. refractor of the Observatoire du Pic-du-Midi (Manchester Lunar Programme).

age of the adjacent mare ground. If we accept the identification of a material of this age with that of a Copernican ray (thrown out obviously at a time when the crater was formed), it would follow that this crater itself must be 800 million years old – and that, therefore, its origin goes back well into archaic (pre-Cambrian) times of the geological history of the Earth. The craters Kepler or Aristarchus in Oceanus Procellarum may be distinctly younger (as suggested by a higher reflectivity of the apron of their ejecta), and Tycho (Figure 8.4) may be of age comparable with that

Fig. 8.5. A photograph, at high Sun, of the craters Theophilus, Cyrillus and Catharina (i.e., the same group as seen at sunset on Figure 8.1). Note that the bright rays diverging from Theophilus show only very faintly on the surrounding mare plains on a photograph taken on 3 August 1965 with the 74-in. reflector of the Helwan Observatory at Kottamia, Egypt (Manchester Lunar Programme).

of Copernicus. On the other hand, Theophilus (Figure 8.5) or Eratosthenes – whose rays have by now been almost obliterated by age – are distinctly older; probably between one and two billion years.

However, perhaps the most interesting result which has emerged from radioactive dating of lunar rocks so far is the fact that – in contrast to larger lumps of crystalline rocks – the soil ('fines') collected at each respective site was found to exhibit ages which were not only substantially higher than those of the larger lumps of rock, but this age turned out to be also closely the same for all localities – and close to 4.6 billion years (cf. Papanastassiou and Wasserburg, 1971). This latter age is, in turn, virtually identical with that of the oldest known meteoritic material intercepted by the Earth; and the presumption is likely that it coincides with the time which elapsed since the solar system was formed.

The lunar soil has, therefore, provided us with a direct evidence that solid matter existed on the lunar surface as far back as 4.6 billion years ago, which has not been re-melted since that time. A schematic picture of this chronology is shown on the accompanying Figure 8.6 (after Toksöz and Solomon, 1973). Nothing of the age comparable to that of the lunar 'fines' has certainly been found anywhere on the Earth. A comparison of the early chronology of the Moon and the Earth is shown on Figure 8.7 after Gast (private information). The reader may note that (with possible excep-

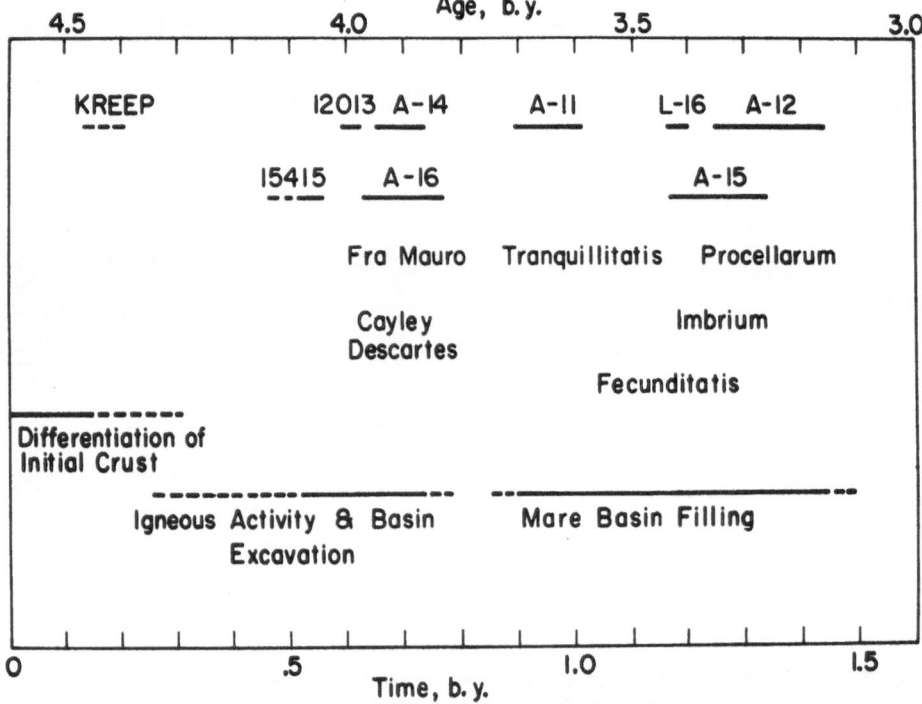

Fig. 8.6. A chronology of the igneous activity on the lunar surface in the first 1.5 billion years (after Toksöz and Solomon, 1973).

tion of certain rocks found recently in Greenland, whose age may approach 4 billion years) the oldest extant pages of the terrestrial 'book of the hours' go back only to 3.5 billion years ago – their ledger commencing just about at a time when the lunar history seems to have come to a standstill. In other words, the first billion years of the Earth's history represents a veritable 'dark aeon' in the life of our planet, on which we find no testimony engraved in the terrestrial stony strata. Since 1969 we have, however, discovered on the Moon a source of material which suddenly illuminated for us the earliest chapter in the history of the Earth-Moon system, and enabled us to reconstruct an almost uninterrupted story of what had been going on in the inner precincts of the solar system since the days of its formation.

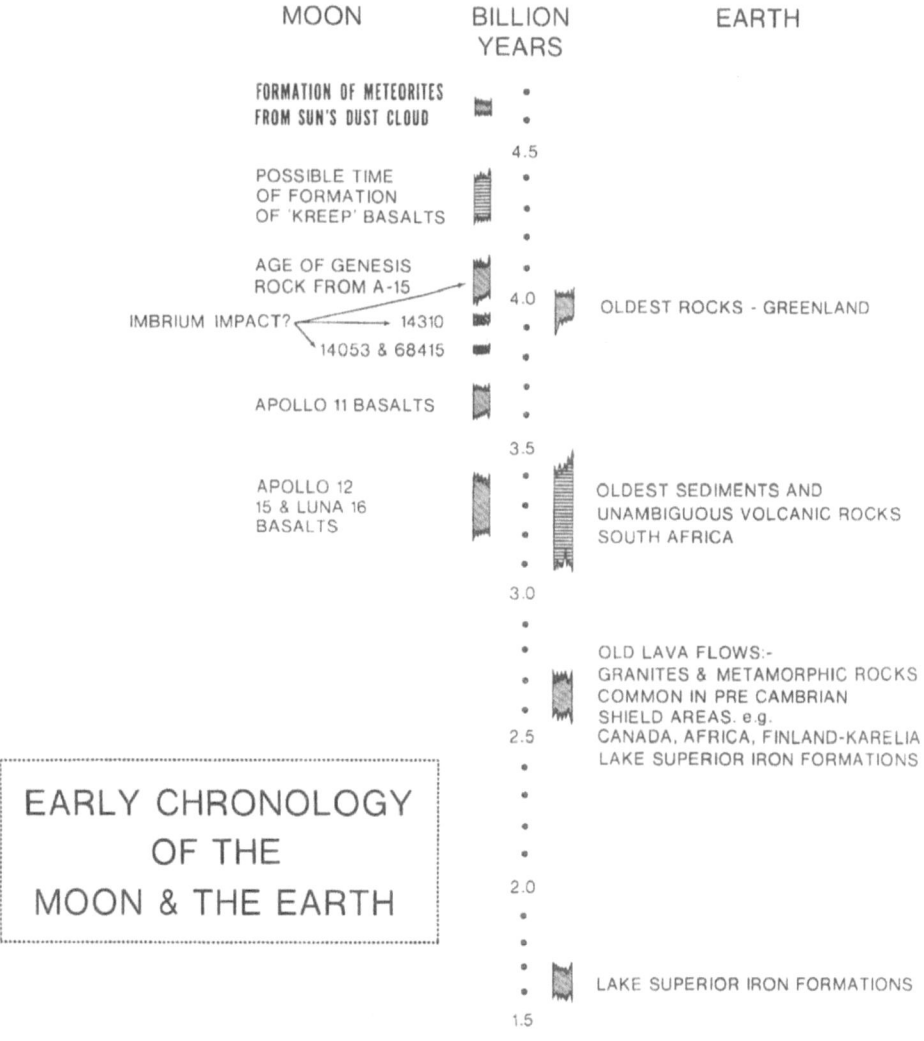

Fig. 8.7. A comparative chronology of the Earth and the Moon (after P. W. Gast, private information).

An account of this reconstruction is being deferred till the concluding Chapter 10 of this book. For the present we wish to note that, in contrast with the Moon, our mother Earth exhibits to the outside world a cosmic face of almost eternal youth – rejuvenated as it is continuously by geological processes such as erosion and denudation of its land by joint action of air and water; or (more important) by continuous continental drifts operative in its mantle and driven by the internal heat engine of the Earth. Very few parts of exposed terrestrial continents, or even ocean floors, are now known to be older than a few hundred million years. In contrast – the Moon (on account of its small mass and heat capacity) can afford none of these means of cosmic cosmetics to make up continuously her face. The latter mirrors, therefore, truly the ages gone by and preserves a reflection of events that occurred long before our own terrestrial continents were formed; and long before the first manifestation of life on Earth flickered in its shallow waters. As a monument of the past, the Moon constitutes one of the most important fossil of the solar system; and a correct interpretation of the hieroglyphs engraved by Nature on its stony face holds indeed a rich scientific prize.

References

Eberhardt, P., Geiss, J., Grögler, N., and Stettler, A.: 1973, *Moon* **8**, 104.

Hartmann, W. K.: 1970, *Icarus* **13**, 299.

Papanastassiou, D. A. and Wasserburg, G. J.: 1971, *Earth Planetary Sci. Letters* **11**, 37.

Schaefer, O. A. and Husain, L.: 1974, in *Lunar Science* V, Pergamon Press, Inc., in press.

Shoemaker, E. M. and Hackman, R. J.: 1962, in Z. Kopal and Z. K. Mikhailov (eds.), 'The Moon', *IAU Symp.* **14**, Acad. Press, New York and London, pp. 289–300.

Toksöz, N. and Solomon, S. C.: 1973, *Moon* **7**, 251.

CHAPTER 9

LUNAR EXOSPHERE

The relative smallness of the mass of the Moon and weakness of its gravitational field, discussed already in Chapter 3, entail many important consequences, and perhaps the most important is the well-nigh complete *absence of any atmosphere* around the Moon which could protect its surface from a direct contact with the conditions prevalent in interplanetary space.

Why should a self-gravitating body of planetary size possess any atmosphere at all? For the terrestrial planets (of masses comparable with those of the Earth) their present atmospheres constitute mixtures of primordial gases with those liberated by thermal de-gassing of their interiors. On the other hand, astronomical bodies as small as the Moon could not have permanently retained any primordial gas; so that such atmospheres as they could possess would have to be regenerated, or accreted.

In order to explain why this should be so, let us recall that the continued existence of an atmosphere around any celestial body, and its composition, testifies to the extent of a stalemate between two opposing tendencies: the attraction of the central body which weighs on each gas molecule in the same way as on any macroscopic object, and will prevent the escape of all those whose velocity v is less than the parabolic velocity $v_{esc} = \sqrt{2Gm_{\text{(}}/r_{\text{(}}}$, where $m_{\text{(}}$ and $r_{\text{(}}$ continue to denote the Moon's mass and radius, respectively; while, on the other hand, the heat pumped in our gas by the Sun (as well as by the surface of the respective planet) maintains the kinetic energy of the gas particles and thus keep the atmosphere distended.

In general, the frequency distribution N of molecules in a gas at a temperature T is governed by the well-known Maxwellian law

$$N(v) = \left(\frac{mv^2}{2\pi kT}\right)^{3/2} \exp\left(-\frac{mv^2}{2kT}\right),\tag{9.1}$$

where m denotes the mass of the respective molecules, and $k = 1.3805 \times 10^{-16}$ erg deg^{-1} stands for the Boltzmann constant. Only such fraction of the total number of particles will be able to escape for which

$$v^2 \geqslant \frac{2Gm_{\text{(}}}{r_{\text{(}}} = 2gr_{\text{(}},\tag{9.2}$$

but since the mean value of v^2 results from (9.1) as

$$\bar{v}^2 = \frac{3kT}{m},\tag{9.3}$$

it follows that the balance of escape will depend on the particular values of $r_{\text{(}}$ and

g as well as T and m. If, in particular, t denotes the time during which the density of an (isothermal) atmosphere diminishes to $1/e = 37\%$ of its initial value is equal (cf., e.g., Spitzer, 1949) to

$$t = \frac{\sqrt{6\pi\bar{v}^2}}{3g} \frac{e^Y}{Y}, \tag{9.4}$$

where

$$Y = \frac{3}{2}\left(\frac{v_{esc}}{\bar{v}}\right)^2. \tag{9.5}$$

When these results are applied to the Moon, for which $g = 167$ cm s^{-2} and $v_{esc} = 2.38$ km s^{-1}, and where the temperature T oscillates approximately between 400 K at noon-time and 100 K during the night, we find that the lightest gases – hydrogen and helium – should disappear from the lunar environment almost immediately. For a hypothetical atmosphere consisting of atomic hydrogen the time t as defined by Equation (9.4) turns out to be approximately 125 min in daytime, and about 215 min (3.6 hr) during the lunar night. A helium atmosphere would similarly dissipate in some 3.6 hr of daylight, or 1.4 yr at night. Atomic oxygen or water vapour already take years for dispersal at daytime, and 10^6 yr at continuing night-time conditions. Molecular oxygen (O_2) could outlast in lunar daylight millions of years; and CO_2 still longer. But the Moon's age of 4.5×10^9 yr is so great that none of the common gases could survive exposure to sunlight and remain attached to it by more than a fraction of its age; the substantially longer lifetimes at night are not really relevant in this connection.

Besides, some of those just mentioned (oxygen, for instance) are so reactive that they would not stay long in free state anyway if in contact with a solid surface, but would form compounds in the surface layer of solid rocks. In all, the rate of dissipation into space (or the formation of solid compounds on the surface) of all but the heaviest (or inert) gases – which are again cosmically very scarce – is so high on the Moon that we should not expect to find any appreciable permanent atmosphere around it; and this expectation has indeed been borne out by all aspects of the observational evidence available to us so far.

The virtual absence of any atmosphere around the Moon was suggested to the early telescopic students of the Moon by the sharpness of the appearance of its face, and by a complete lack of any cloud cover. Stark shadows cast by the mountains are not mitigated by atmospheric light scattering. Moreover, no evidence of *refraction* has ever been noted at the Moon's limb, whenever our satellite places itself between us and any more distant celestial object.

This absence of refraction constitutes by itself a sufficiently stringent test to enable us to conclude that the density of a hypothetical lunar atmosphere above the surface must be less than one part in ten thousand of the terrestrial air density at sea level. In more recent years, this upper limit has further been lowered by repeated but so far fruitless quests for an indication of *twilight phenomena*, which should be produced in a hypothetical lunar atmosphere during sunrise or sunset.

In order to appreciate more fully the power of this method, consider the light reaching us, say, from any spot near the central part of the apparent lunar disc at the time of the first quarter just before sunrise, when the ground is still immersed in darkness but the space above it already receives the first rays of the rising Sun. This light should, in general, consist of three parts:

(1) the light from the lunar surface as illuminated by the Earth;

(2) the light of the Sun scattered in the direction of the line of sight in a hypothetical lunar atmosphere; and

(3) the moonlight scattered in the terrestrial atmosphere in the direction of the observer.

In the efforts to separate the component (2) – which alone bears on the presence of any gas around the Moon – from the parasitic (but much more intense) diffuse light arising from the sources (1) and (3), Nature lends us fortunately a helpful hand: for the light scattered on gas in a direction which is perpendicular to that of its original incidence should be noticeably polarised, while the diffuse light arising from the lunar as well as the sky background should be essentially free from polarisation. A search for the manifestation of a hypothetical lunar atmosphere reduces, therefore, to the detection of a polarised component in the diffuse light reaching us from an element of surface area of the Moon just before sunrise near the first quarter.

Actual measurements of this type were first performed with sufficient accuracy by Fessenkov, whose results were published in 1943. The outcome was again negative; no trace of polarisation was detected; and the precision of the experiment was such as to lead Fessenkov to conclude that the density of a hypothetical lunar atmosphere – if any – must be less than one part in a million (or 10^{-6}) of that of the terrestrial atmosphere at sea level. If, incidentally, a hypothetical lunar atmosphere were as dense as 10^{-4} of the terrestrial one, the twilight zone illuminated by the Sun at the time of the new Moon would endow the lunar disc with an aureola whose light would, in fact, be more intense than the earthlight on the dark side of the Moon and thus be readily detectable – which it certainly is not.

In more recent years, this limit has been further lowered by Bernard Lyot, who, by examining the light reaching us from the regions just beyond the cusps of a lunar crescent eliminated the effect of earthlight represented by (1). By making his observations from the lofty height of the Observatory on Pic-du-Midi, in the French Pyrénées, through clear atmosphere renowned for its superlative seeing, and using the coronograph (a telescope specially designed by him to minimize the amount of light scattered in the instrument itself), Lyot was in a position to detect much smaller traces of polarisation in the light of the cusps of the Moon than Fessenkov was able to do near the centre of the lunar disc. However, in spite of this greatly increased instrumental precision, all Lyot's results were again purely negative: no trace of polarisation was detected; and from its absence Lyot concluded in 1949 that the density of any hypothetical lunar atmosphere must be less than 10^{-8} of that of the terrestrial one. Moreover, after Lyot's death in 1952 his work was continued by Dollfus, who by resorting to further refinements depressed the possible upper limit for the density of a hypothetical lunar atmosphere to 6×10^{-10} of the terrestrial one (1956). If the Moon possesses any gaseous atmosphere, its density above the surface cannot, therefore, exceed some 7×10^{-13} g cm^{-3}; but how much less the actual density may be this method cannot tell.

The most sensitive Earth-based test for the presence of a hypothetical lunar atmosphere can be attempted by a resort to observations in the radio (rather than optical) domain of the electromagnetic spectrum: namely, to search for the effects of refraction exerted by a hypothetical lunar ionosphere on radio stars (or signals from a spacecraft) occulted by the Moon. As is well known (cf., e.g., Link, 1956), the effects of such a refraction is to prolong the duration of such occultations by time increments which can be measured, and converted into the electron density prevalent above the lunar surface. Past observations of this phenomenon (cf. Costain *et al.*, 1956; Elsmore, 1957) disclosed that the electron density along a line of sight clearing the lunar surface cannot exceed some 10^4 free electrons per cc. Although this value is not directly convertible into a density in g cm^{-3} because of the unknown degree of ionization prevalent in the lunar atmosphere, it is indicative of a density even lower than 10^{-12} g cm^{-3}.

A gas density of this order would represent a pretty hard vacuum from the point of view of the terrestrial physicist – and one which is, incidentally, attained in our own atmosphere at an altitude of approximately 180 km above sea level. However, even at such great heights, the number of gas particles remains still of the order of 10^{10} per cm^3; and although the mean free path of such particles between mutual collisions is of the order of 100 m, even so rarefied a gas can manifest itself in different observable ways. It would not, to be sure, offer any protection to the surface beneath from impinging meteorites. A hypothetical lunar atmosphere of surface density of the order of 10^{-12} g cm^{-3} would not decelerate any meteoritic material – small or large – enough to cushion significantly its impact. All solid particles in space intercepted by the Moon must hit its surface essentially with their cosmic velocities, and spend themselves on the surface rather than in their passage through the atmosphere as is the case on the Earth. Needless to say, the rate of meteoritic matter (of all sizes) per unit area of the Earth as well as of the Moon should be very approximately the same. In quest of its observational verification on the Moon we should not, however, look for any luminous trails during approach; but rather for instantaneous flashes of light which could be produced by impact of a meteorite on solid rocks.

However – returning to our own atmosphere at an altitude of 180 km above sea-level – even though it is insufficient to affect the velocity of the meteors traversing through it, it can give rise to other interesting phenomena. Between 180–200 km above sea level we would find ourselves in the midst of the auroral zone, where luminescent gas stimulated by the impact of corpuscular sunrays produces the beautiful displays of 'northern lights'. Are there similar aurorae on the Moon? Herzberg (1946) pointed out in 1946 that a search for emission spectra of such displays around the bright limb of the Moon might constitute one of the most sensitive tests of the presence of a hypothetical atmosphere of our satellite. No trace of any such emission has, however, been detected until the advent of the last two Apollo missions, on which we shall have more to say later on. As long as telescopic observations at a distance remained our sole source on information on the lunar environment, it seemed that the Moon could not possess any atmosphere of density exceeding 10^{-12} g cm^{-3} at its surface; but how far the actual density could lie below this limit remained conjectural. Even the upper limit just stated appeared, however, to be rather uncomfortably low – for several reasons.

In order to explain why, let us recall that, on account of the feeble lunar attraction, all light gases would escape indeed completely from its gravitational field in the course of the Moon's long astronomical past; and most of the heavier gases (such as SO_2, for instance) would react again with surface rocks to form solid compounds. This should, however, not apply to the 'inert' gases like argon, krypton, or xenon, which do not form compounds, and which are sufficiently heavy for their rate of escape from the gravitational field of the Moon to be moderately low (the mean atomic velocities of thermal agitation of these gases at 0 °C being 414 m s^{-1} for Ar, 287 m s^{-1} for Kr and 229 m s^{-1} for Xe).

The case of argon is particularly arresting, because – quite apart from any aboriginal amount of this element which the Moon may have retained from primordial times – its supply must be continually replenished by radioactive decay of the heavy isotope of potassium (K^{40}). The total mass of the Moon of 7.35×10^{25} g, should then contain about 8.8×10^{22} g of potassium; and of this about 9.7×10^{17} g should be the radioactive K^{40} decaying (by β-disintegration) into the common isotope of argon (Ar^{40}).

The total disintegration of all lunar K^{40} should, therefore, create 9.7×10^{17} g or 1.5×10^{40} atoms of argon – as compared with some 10^{44} gas particles now constituting the terrestrial atmosphere. Just how much of this argon managed to escape to the surface from the lunar mass by gradual degassing (caused by the rising internal temperature) remains conjectural within fairly wide limits. However, on almost any reasonable guess (coupled with the known rate of escape of argon from the gravitational field of the Moon) the amount of argon in the lunar atmosphere should add up to more than the upper limit of the lunar air density consistent with the absence of perceptible twilight phenomena.

When we come to consider the remaining two heavier inert gases – krypton and xenon – the situation becomes even more embarrassing. These gases likewise do not form compounds, and their atomic weights (two and three times that of argon) are such that their gravitational escape from the Moon (by collisions) becomes effectively impossible (10^{24} yr for Kr and 10^{41} yr for Xe) even if exposed to noon-time conditions for the entire age of the Moon. Moreover, even if no krypton and xenon were originally present around the Moon, more than enough of them would have been produced in the lunar crust by a spontaneous disintegration of the heavy isotope of uranium (U^{238}), by interaction of the lighter isotope U^{235} with neutrons produced on the lunar surface by impact of cosmic rays, and also (for xenon) by spontaneous decay of the radioactive isotope of iodine (I^{129}).

If uranium is present in average lunar rocks in the percentage indicated by the Apollo results, it follows from known characteristics of uranium disintegration that 2.5×10^{24} atoms of xenon, and 1.5×10^{23} atoms of krypton should have been produced in the lunar surface layer 1 cm thick in the course of 4×10^9 yr. The liberation of a gas so formed into the atmosphere could, moreover, occur whenever the solid rocks are mechanically disturbed by some event (such as the impact of a meteorite, or tectonic processes operating in the crust).

According to the present estimates of the frequency of meteor impacts on the Moon, during 4×10^9 yr its surface should have been effectively 'disturbed' by such impacts down to a depth of 1–10 km. If so, the cumulative effect of it would have been a release of 10^{34}–10^{35} atoms of krypton and 10^{32}–10^{33} atoms of xenon, constituting a permanent atmosphere of 10^{-9}–10^{-10} of the terrestrial air pressure on the ground (and consisting of approximately 93% of Kr and 7% of Xe).

Why is this gas not there? Several authors (Edwards and Borst, 1958; Herring and Licht, 1959, 1960) pointed out, some years ago, that atoms of heavy gases can be mechanically 'blown off' the Moon into space by collision with corpuscular radiation (mainly protons) emitted continuously – and, sometimes, in angry energetic 'puffs' – by our Sun. The existence and intensity of this 'solar wind' and its occasional gusts usually associated with flares and other sudden disturbances of the solar surface, is now well known from the work of deep-space probes, and indirectly attested by such terrestrial phenomena as polar aurorae and magnetic storms. The knocking-off power of this solar wind is, in fact, sufficient to remove most of the heavy inert gases from the lunar surface, and thus despoil it even of such scanty vestiges of gas envelope which its own feeble gravitational attraction would enable it to retain.

Subsequently, Öpik and Singer (1960, 1961) or Hinton and Taeusch (1964) pointed out another and more effective way of gas removal from the lunar surface: namely, by ionization. As long as gas remains neutral, its atoms or molecules can be removed only by collisions – be it with other molecules, or particles of the 'solar wind'. Should, however, the gas particles become ionized and acquire thus positive electric charge, the positive electrostatic charge of the sunlit hemisphere of the Moon, acquired by photoionization of the lighter elements in its crust, can remove ions by repulsion far more effectively than could be accomplished by the collisions with the protons of the solar wind. It seems, therefore, possible to explain now the well-nigh complete absence of free gases around the Moon in more than one way, and thus reconcile theoretical expectations with the observed facts to a complete satisfaction of all investigators.

Such gases as are left by these processes to cling to the lunar surface for a limited time do not constitute any real atmosphere – in which individual gas particles are balanced up by mutual collisions – but rather a transient *exosphere*, in which the individual atoms or ions describe essentially free-flight trajectories in the prevailing gravitational or electrostatic field. Each planetary atmosphere is bound to peter out into such an exosphere on its outer fringe bordering on interplanetary space; but on the Moon this exosphere apparently reaches down to the solid surface itself. It is this surface which effectively controls the 'temperature' of the exosphere, whose particles unimpeded by collisions travel to and fro in ballistic trajectories whose typical time of flight is of the order of a few hundred seconds – after which they will collide again with the surface to be re-heated and bounced back. Eventually, such particles will be ionized by solar radiation, and escape from the Moon by spiralling about the magnetic lines of force carried by the solar wind – unless they collide promptly with the lunar surface to become neutralized by electron capture.

For example, many tons of gas fuel of relatively high molecular weight which have been burned in the lunar environment in the course of successive Apollo landings and take-offs are being held thus by the lunar surface. On the other hand, all hydrogen atoms with the average velocity of 2.9 km s^{-1} acquired from lunar noon-time surface at 400 K will escape into space; only cooler atoms can return to form a hydrogen exosphere of some 3000 km scale height – the latter being comparable with the diameter of the Moon as a whole. If, however, we turn to the heaviest inert gas – xenon – in lunar environment, a 400 K surface should impart to its atoms a mean velocity of only some 0.23 km s^{-1} – well below that of escape from the gravitational field of the Moon. Free xenon around the Moon should, therefore, constitute a much more compact exosphere of scale-height close to only 20 km.

How many atoms are actually populating the lunar exosphere at the present time? We mentioned already in this chapter that optical phenomena observable from the Earth rule out the presence, around the Moon, of any envelope whose density would exceed 10^{-12} g cm^{-3}. On the other hand, we know that the solar wind is continuously transporting to the Moon a small amount of gases escaping from the solar atmosphere, whose average contributions are listed in Table 9-1 (after Vondrak *et al.*, 1974). Thus – regardless of any de-gassing of the lunar interior – the arrested solar

TABLE 9-1

Gas acquisition by the Moon
through solar wind

Element	Rate of influx (in g s^{-1})
H	40
He^4	8
N^{14}	0.03
O^{16}	0.2
Ne^{20}	0.07
Si^{28}	0.05
Ar^{38}	0.004

wind alone should be sufficient to endow the Moon with an exosphere of density of the order of 10^{-22} g cm^{-3} at the time of the 'quiet' Sun, and as much as 10^{-20} g cm^{-3} during major flares. Where in between these limits the actual density may happen to lie we did not learn until the advent of spacecraft.

The first contribution to our knowledge in this respect was made by lunar orbiting satellites, the characteristics of which have been listed in Table 1-3 on p. 17. Collectively, these satellites completed more than 10000 revolutions around the Moon, in orbits of various sizes and eccentricities. If any appreciable exosphere existed at an altitude of their periselenia passages (see column 4 of Table 1-3) a drag exerted by it on moving vehicles would have led to a secular loss of kinetic energy, manifesting itself through diminishing semi-major axis (or the orbital period) of the spacecraft. A careful analysis of existing observations failed to detect any secular period change which could be indicative of air drag; and the upper limit above which period changes would have been detected, has relegated the density of a hypothetical lunar atmosphere at 50 km above ground to less than 10^{-15} g cm^{-3}.

Further refinements of this had to await the outcome of the Apollo missions. Up to the end of 1970, the best information we had was obtained with the aid of cold cathode pressure gauges mounted on the lunar surface by Apollo 12 and 14 missions (cf. Johnson, 1971). These measurements indicated that the maximum lunar atmospheric density over the daylight hemisphere appeared to correspond to some 2×10^5 particles per cc at night, increasing about 100 times during the daytime. The low night-time values indicated that many kinds of gas (e.g., contaminants injected into the lunar environment by spacecraft) freeze out a temperatures of 100 K or less prevalent during that time; and their reappearance at sunrise is no doubt due to the release of such gases when the temperature increases.

In the course of Apollo 16 mission in April 1972, Carruthers and Page (1972) operated on the Moon a far-UV photographic camera, which recorded (on some photographs) a faint glare in the light of hydrogen Lα-line, the intensity of which increased towards the lunar horizon. The experimenters interpreted this phenomenon as hydrogen emission from the lunar exosphere; but published so far no quantitative data on its density.

Moreover, during the last Apollo 17 mission in December 1972, Fastie and his collaborators operated a far UV spectrometer in orbit around the Moon (cf. Fastie, 1973), in quest of spectroscopic indications of a gaseous envelope around the Moon in approximately the same spectral domain as Carruthers and Page. The experiment performed satisfactorily throughout the Apollo 17 mission, but its results were largely negative. None of the expected gases (such as krypton or xenon, oxygen or nitrogen) were found to be spectroscopically present – except for hydrogen; but the latter only in a trace amounting to less than 10% of the concentration expected from the solar wind alone.

While the definitive results of these experiments are yet to be published, there is no doubt already now that the lunar environment is extraordinarily devoid of any free gas. They prove that the Moon is *not* degassing now to any appreciable extent – anywhere nearly at the same rate as our Earth – nor was it for times comparable with that of escape of the respective gases from the gravitational field of our satellite. The well-known spectroscopic observations of transient luminous phenomena in the crater Alphonsus by Kozyrev in 1958 (described in Chapter 22 of *Moon II*) could not, therefore, have been produced by gas whose photodissociation would have produced the alleged Swann bands of the C_2-molecule. The reason is the fact that the bulk of gas of such molecular weight would still have to be present around the Moon at the time of the Apollo missions, and be detected by the suprathermal ion devices of their ALSEP packages. A completely negative result of such measurements strengthens the surmise that what Kozyrev observed was no gas emission, but ground luminescence of solid state. The solar wind alone transports more gas to the Moon than our satellite can retain – a fact testifying to the effectiveness of gas removal by ionization. The total amount of hydrogen in the lunar exosphere now does not seem to exceed 1 kg of free gas; and that of xenon, about 1 ton.

Such amounts are, of course, very small in comparison with gas released in recent years in lunar proximity by human action. The soft-landers and orbiters exported between 1966–1967 to the Moon an amount of gas which could already then have altered appreciably the chemical composition of the lunar exosphere; and the masses of gas released around the Moon by successive Apollo missions were apt profoundly to alter the composition. On the injection of the command module in the lunar orbit (cf. Chapter 2) about 10 ton of exhaust gas are released over the far side of the Moon; but as – on braking – the combination of the vehicle velocity and gas exhaust velocity exceeds that of escape from the gravitational field of the Moon, most of this gas was lost into interplanetary space. During the descent, about $7\frac{1}{2}$ ton of additional exhaust gas are released; and most part of this mass will remain on the Moon. On the first leg of the return journey, about 5 ton of exhaust gases are ejected again mainly over the Moon's far side; but since – this time – spacecraft and exhaust velocities are in opposite directions, most of this gas should likewise be retained by the Moon.

If so, however, then each one of the seven Apollo missions should have left on the Moon about 10 ton of exhaust gases alone – and the total contribution of all could not have been much less than 100 ton. This amount is very large in comparison with the natural gas around the Moon; and if spread uniformly around the Moon it should produce a gaseous envelope of density close to 10^{-10} g cm^{-3} at an altitude of 100 m above the lunar surface (with a mean-free-path of the individual particles close to 1 m). Moreover, the molecular weight of the constituent particles (estimated to consist of some 36% of H_2O, 32% of N_2, 13% of H_2, 9.6% of CO, 3.7% of CO_2, 1.9% of H and 1.6% of OH; cf. Aronovitz et al., 1968) is sufficiently high to make the rate of a gravitational dispersal of most of these molecules very long in comparison with the time interval between successive Apollo missions.

Why is this gas not there? Neither the gravitational, nor electrostatic rate of removal is sufficiently fast to explain a well-nigh complete absence of such gas around the Moon, established by observations described earlier in this chapter. We are, therefore, driven to a conclusion that such gases must be very rapidly *absorbed* by the lunar surface; with the possibility of occasional release triggered by local events. The sudden (though temporary) release of water molecules over a 14-hr interval on 1971 March 7, recorded by the Apollo 14 suprathermal ion detector (cf. Freeman et al., 1973), may have been caused by some such event of as yet unspecified nature.

The extreme tenuity of the lunar exosphere at the present time is – we repeat –

now a fact sufficiently attested by observations and experiments carried out in the lunar environment itself. This fact would, in turn, guarantee a well-nigh complete absence of erosion which could be caused on the Moon by any kind of air (or liquid). Has it been so always throughout the long astronomical past of our satellite? Before giving an affirmative answer we wish to mention, in conclusion of the present chapter, one possibility which could, at times, have temporarily altered such a situation: namely, whenever in the past the Moon collided in space with the nucleus of a *comet*.

As is well-known, the nucleus of a comet – the only part of their anatomy possessing any kind of permanence – constitutes an iceberg of frozen hydrocarbons and other compounds of moderate molecular weight, which remain in solid state as long as the comet floats freely in space sufficiently far from the Sun; and whose gradual evaporation in more moderate zones of interplanetary climate gives rise to the beautiful though ephemeral phenomena of cometary comas and tails. When, however, a cometary nucleus strikes a solid obstacle – such as the surface of the Moon – a conversion of ice to gas is virtually instantaneous. Inasmuch as cometary impacts on the Moon must have occurred many times during the long astronomical past of our satellite, the question naturally arises as to the fate of the gas which must have been let loose over the lunar surface each time when a comet committed suicide in this manner.

The total amount of gas which can be acquired by the Moon by such catastrophic encounters is far from negligible. According to the data contained in Richter's recent monograph (cf. Richter, 1962), the average mass of a cometary nucleus is of the order of 10^{18} g (i.e., 10^{10} times as large as the total amount of gas released around the Moon by all spacecraft between 1966–1972); and comets that may have been 10 or even 100 times as massive are a part of historical records (the most recent being comet Arend-Roland 1956h). In order to place these values in proper perspective, let us recall that the total mass of our terrestrial atmosphere (generated essentially by de-gassing of the Earth's interior) is close to 5.1×10^{21} g – i.e., about a thousand times that of an average comet; and 3×10^{20} g would thus be sufficient to provide a column of gas of equal mass above each unit area of the lunar surface as we have on the Earth. Therefore, a comet of an average mass could, on impact, provide the Moon with a gaseous envelope containing about one per cent of the terrestrial air mass above each unit area of the surface; and a heavier comet could import proportionally more.

What is the chemical composition of gas that could be acquired by comet decay? An analysis of the cometary spectra (cf., e.g., Swings and Haser, 1956) disclosed the presence of a considerable number of molecular constituents, of which those with molecular weight between 20 and 50 are listed in the accompanying Table 10-1. No one knows for sure the pristine composition of the nucleus which is the origin of all gases in the coma or the tails; but since many of these could have originated by photo-dissociation of more complicated parent molecules present in the nuclear material, the latter may well contain many constituents of molecular weight well in excess of 50.

We do not know, of course, what fraction of the cometary gases liberated by sudden melting of the nucleus will dissipate thermally into space on impact; but such fraction of them as may be allowed to cool enough to reach thermal equilibrium with the lunar surface would then be vouchsafed – by Equation (9.4) – half-lives listed in column 3 of Table 9-2, corresponding to lunar noon-time conditions ($T = 400$ K).

TABLE 9-2

Dispersal rate of neutral cometary
atmospheres on the Moon

Molecule	Molecular weight	Log of half-life t (in yr) for $T = 400$ K
H_2O	18	0.83
C_2	24	2.88
CN	26	3.54
CO^a	28	4.22
N_2^a	28	4.22
C_3	36	7.04
CO_2^a	44	9.83

[a] These gases appear as ionized in cometary tails; but in the nuclei are undoubtedly present in neutral state.

A glance at the data compiled in Table 9-2 reveals that, while constituents of molecular weights close to 25 could remain gravitationally attached to the Moon for time intervals of the order of 10^3 yr of daytime (and almost indefinitely long at night), those of molecular weight of 40–50 could remain so attached for 10^8–10^9 yr! Time-intervals of this order are comparable with the total age of the Moon (cf. Chapter 8), and certainly long in comparison with an average time-lapse between successive cometary impacts on its surface. Since, moreover, each such impact could have provided the Moon with an atmosphere of mass of the order of one per cent of the terrestrial air mass above unit area, why is it not there?

Optical observations described earlier in this chapter leave no room for doubt that the actual amount of gas around the Moon now must be less by several orders of magnitude to escape detection. On the other hand, it is equally certain that comets must occasionally strike the Moon; and the only way to reconcile this with the apparent absence of gas on the Moon now is to admit that cometary gas is being removed from lunar environment *faster* than predicted by the kinetic theory of neutral gases. The principal mechanism of removal may again be absorption by the lunar surface; though absorption of so large an amount of gas should leave a chemical imprint in its composition. No such imprint is evident in the chemical composition of lunar material brought back from any locality on the Moon so far – in particular (cf. Chapter 7), carbon or its compounds are deficient, rather than enhanced, in their relative abundance as compared with the Earth. It is, of course, possible that enhancement may be limited to regions of actual cometary impacts, and that none of our spacecraft happened so far to land in the proximity of one.

But be so as it may, cometary impacts must have occurred repeatedly on the Moon in the course of its long astronomical past, and imparted large amounts of hydro-carbons – large enough to provide for a much denser atmosphere around the Moon than the one which we observe today. Such spells must, however, have been transient – ending in complete dispersal over time short in comparison with the average interval between successive impacts. But, during such spells, the surface of the Moon could have been temporarily exposed to aeolic erosion; or – possibly – even to the effects of flowing water. With so many comets around us in interplanetary space, it is thus impossible to assert categorically any longer than it never rained on the Moon, or that its landscape was never swept by winds. We wish to reiterate, however, that with such evidence as we now possess the role of these processes in shaping up the lunar surface – if not altogether negligible – must have been at best highly localized and very small indeed.

References

Aronovitz, L., Koch, J., Scanlon, J. H., and Sidran, M.: 1968, *J. Geophys. Res.* **73**, 3231.

Carruthers, G. R. and Page, T. L.: 1972, in *Apollo 16 Prelim. Sci. Rept.*, NASA SP-315, 13-1.

Costain, C. H., Elsmore, B., and Whitfield, G. R.: 1956, *Monthly Notices Roy. Astron. Soc.* **116**, 380.

Dollfus, A.: 1956, *Ann. Astrophys.* **19**, 71.

Edwards, W. F. and Borst, L. B.: 1958, *Science* **127**, 325.

Elsmore, B. and Whitfield, G. R.: 1955, *Nature* **176**, 457.

Fastie, W. G.: 1973, *Moon* **7**, 49.

Fessenkov, V. G.: 1943, *Astr. Zhurn.* **20**, 212.

Freeman, J. W., Hills, H. K., Lindeman, R. A., and Vondrak, R. R.: 1973, *Moon* **8**, 115.

Herring, J. R. and Licht, A. L.: 1959, *Science* **130**, 266.

Herring, J. R. and Licht, A. L.: 1960, in H. Kallmann-Bijl (ed.), *Space Research* I, North-Holland Publ. Co., Amsterdam, pp. 1132–1145.

Herzberg, G.: 1946, *Pop. Astron.* **54**, 414.

Hinton, F. L. and Taeusch, D. R.: 1964, *J. Geophys. Res.* **69**, 1341.

Johnson, F. S.: 1971, *Rev. Geophys. Space Phys.* **9**, 813.

Link, F.: 1956, *Bull. Astron. Inst. Czech.* **7**, 1.

Öpik, E. J. and Singer, S. F.: 1960, *J. Geophys. Res.* **65**, 3065.

Öpik, E. J. and Singer, S. F.: 1961, *Science* **133**, 1419.

Richter, N.: 1962, *Physics of the Comets*, Methuen and Co., London.

Spitzer, L.: 1949, in G. P. Kuiper (ed.), *The Atmospheres of the Earth and Planets*, Univ. of Chicago Press, Chicago; Chapter VII.

Swings, P. and Haser, L.: 1956, *Atlas of Representative Cometary Spectra*, Centeric Press, Louvain.

Vondrak, R. R., Freeman, J. W., and Lindeman, R. A.: 1974, *Lunar Science* V, Pergamon Press, Inc., in press.

CHAPTER 10

ORIGIN AND EVOLUTION OF THE MOON

In the preceding nine chapters of this volume we gave a brief account of the methods of space research of the Moon, as well as of the principal results obtained with their aid. The aim of this concluding chapter will be to attempt a synthesis of this knowledge in so far as it bears on the origin and evolution of our satellite.

The time at which the origin of the Moon is to be sought can now be dated with relative precision; for the radioactive determinations of the age of the lunar soil (i.e., the 'fines' covering apparently most part of the lunar surface) led to closely concordant values between 4.6 and 4.7 billion years (probably closer to the lower limit) for the time which elapsed since their solidification. This age agrees, moreover, quite closely with that established for the oldest meteorites which found their way into the terrestrial laboratories; and are generally regarded to date the origin of the solar system as a whole.

There seems no longer any room for doubt that the formation of the Sun and of its attendant system of lesser bodies from the primordial solar nebula constituted events which were contemporary rather than consecutive; and a close concordance between the radioactive ages of the oldest meteorites and of the lunar soil strongly suggests that our Moon originated also as a by-product of the same process and at the same time. Gone are the days when it was possible to entertain the idea of the Moon having been an 'interstellar tramp' subsequently captured by the Sun and, eventually, the Earth; the age determinations of lunar materials prove that this was not the case.

Secondly, the indications based on the presence of decay products of short-lived radioactive elements in the old lunar soils disclosed that, once that last stage of the formative process of the solar system was reached, the final collapse of the primordial nebula and accumulation of the planetary globes did not take more than a few million years; and our Moon too was no doubt formed within this time.

Third, the chemical evidence we now possess on the composition of the lunar rocks leaves but little room for doubt that the Moon was formed by an accumulation of solid particles at low or moderate temperatures. We know this because an agglomeration of mass comparable with that of the Moon could not have come into being by a condensation of gas at a temperature prevalent in the original solar nebula; for the self-attraction of a mass so small could not withstand (except for the heavy elements) the effects of collisions tending to disperse it. Certainly the volatile elements found in the lunar rocks could not have been retained in their observed proportions if the bulk of the lunar mass condensed directly from gas.

We know, to be sure, still very little of the chemical composition of the deep interior of the Moon; for all rocks in our possession have been collected on the surface, and there is no compelling evidence that these ever emerged from any great subsurface depth. But, on the other hand, we know that the bulk density of lunar matter cannot vary more than a few per cent between the crust of the Moon and its center; suggesting no great difference in composition throughout its interior as well. This uniformity need not be exact; for if the mass of the Moon took a few million years to accumulate, the material accreted in the last stages of this process could already have been depleted in some constituents which may be more abundant in the original core. There is, to be sure, no direct evidence that this was actually the case; but the possibility should be kept in mind.

The question can, moreover, be asked: how is it possible to reconcile the idea of the origin of the Moon in a 'cold' state with the fact that all lunar rocks imported by the Apollo missions – including the soil 4.6 billion years of age – are igneous, showing evidence that they solidified from magma at a temperature of 1100–1200 °C under highly reducing conditions? The magmatic nature of the rocks makes it indeed evident that they all must have been molten at some time; and the problem is only to identify the time and place where this melting occurred.

Many mineralogists who studied these rocks have automatically assumed that the differentiation of lunar rocks must have occurred on the Moon – when these rocks have already been a part of the lunar globe. The idea that the Moon was ever fluid throughout its mass can, however, be dismissed almost out of hand; for it could not have accumulated in this way because of insufficient self-attraction; and if a ready-made solid Moon melted subsequently (as a result of the conversion into heat of collisonal energy of accumulation, for instance, or by an excess concentration of short-lived radioactive elements), it would not have had a chance to cool off in $4\frac{1}{2}$ billion years to the stage at which four-fifths of its mass have solidified sufficiently to transmit transversal seismic waves (cf. Chapter 5).

The cosmochemists engaged in lunar research pointed out, to be sure, that a melting of the lunar crust down to a sub-surface depth of no more than 200 km would be sufficient to account for the observed degree of chemical differentiation of surface rocks from an initially undifferentiated magma. To bring such a shell to the melting point – by heat generated by impacts, or by electromagnetic induction (cf., e.g., Sonett and Mihalov, 1972) is indeed an easier task than to melt the Moon as a whole; and subsequent cooling of the outer shell can again be accomplished within relatively short time. However, while this seems also the simplest assumption to consider, it is necessary to point out at the same time that other important evidence furnished by the Apollo missions – that bearing on the shape and moments of inertia of the lunar globe (cf. Chapter 4) – is seriously at variance with such an assumption.

As we pointed out already on pp. 84–86 of Chapter 4, the shape of the Moon as traced by the laser altimeters of the Apollo 15–17 missions is *not* one in which a global ocean of fluid magma could have solidified at any time, and at any distance from the Earth; for it does not represent a surface of hydrostatic equilibrium in any gravitational field to which our satellite could have been exposed at any time in the past.

One could, perhaps, argue that the present shape of the Moon may have been disfigured from one of equilibrium and abraded by meteoritic bombardment in the first few hundred million years of the Moon's existence (on which more will be said later on in this chapter). However, the moments of inertia (which represent volume, rather than surface, properties of our satellite) are likewise seriously out of harmony with the consequences of hydrostatic equilibrium on the Moon – to an extent discussed already in Chapter 4. And we wish to recall that, in order to account for the observed facts, their anomalies must be located in the outer layers of the Moon's mantle – i.e., precisely where the alleged melting should have taken place. If we were to go by the astronomical evidence alone, we could conclude – plausibly enough – that the present shape of the Moon, and its moments of inertia, are due to the vagaries of its growth in the last stage of accretion.

That this accretion could have led to the formation of a body lacking strict radial symmetry is natural enough; for the converse would indeed have been more surprising in the case of a body which grew up by an accumulation of solid particles. But how to explain then the igneous nature of the rocky layer deposited on the Moon 4.6 billion years ago? One explanation could be to assume that *the rocks and solid debris whose coalescence gave rise to our satellite were differentiated prior to their accretion*; in other words, that *the Moon could have grown out of material already differentiated*.

Such a suggestion does not, unfortunately, lend itself readily for any cosmochemical or petrographic checks. The gravitational field – which would have been very small for differentiation in the pre-lunar past, and virtually the same as that prevalent now on the lunar surface if this is where the crystalline rocks we possess last solidified – does not seem to influence the chemical composition, or crystal structure, of these rocks to any appreciable extent. We also possess still next to no positive information on the 'anatomy' and time-scale of the actual accretion process; and have no idea as to whether the chemical differentiation exhibited by lunar surface rocks is merely skin-deep, or extends into the interior of our satellite; for if the former were true, differentiation could be characteristic only of material accreted in the last stage of the formation of our satellite. We also know that the primordial solar nebula could never have collapsed into the family of bodies to which it gave rise (including the Sun) if it did not contain an appreciable ingredient of solid particles to begin with – but how these were formed, or what their chemical composition and mineralogical structure could have been, we are still largely ignorant. And as long as this is true – which may be for some time – it should be unwise to insist that lunar rocks must have acquired all their observed characteristics after they have become parts of the Moon, and to close our eyes to alternative possibilities.

1. The Earth-Moon System

In the first part of this chapter we have considered out Moon as an independent member of the solar system in its own right, with no regard for the fact (perhaps the best established one of all referred to in this book) that it happens also to be the satellite of our Earth. What is indeed the relation of the Moon to our planet? Is it its foundling or true-born child?

From the early days until the advent of the space age the possibility has been under consideration that the mass of the Moon was once a part of the Earth, and separated from it – at an unspecified time – by a combination of the centrifugal force and tidal attraction of an external body – be this the Sun or (perhaps) some other planetary body. A conjecture was even voiced once (by Osmond Fisher) that the Pacific basin represented a 'scar' in the terrestrial crust where the Moon once separated from the Earth – a view later invalidated by many facts.

Although the hypothesis of the origin of the Moon from the Earth does not lack adherents even at the present time (cf., e.g., Wise, 1963, 1969; O'Keefe, 1969, 1970), it is very difficult to accept for several reasons. First, we know now that the Moon must have originated at the same time as the Earth, and long before the Pacific basin was formed on our planet. Secondly, we know now (cf. Hagihara, 1972) that the Earth–Sun distance has not undergone any appreciable change since the time of the formation of the solar system as a result of planetary perturbations (and scarcely otherwise); so that the tidal pull of the Sun on the Earth was never appreciably different from what it is today. Under such circumstances, the total momentum of the Earth-Moon system should have been very largely *conserved* since the time of its formation up to the present; and its value (based on the present properties of this system) happens to be such that if it were stored entirely in the axial rotation of our planet, the Earth would spin several times faster than at present, but this would *not* make it dynamically unstable.

In other words, the total angular momentum of the Earth-Moon system is *not* sufficient to cause their combined mass to undergo any kind of fission in two parts – let alone in the mass-ratio of 81:1 characteristic of the system. Lastly, an analysis of the chemical composition of the rocks brought back from the Moon in the past few years disclosed (cf. Tables 7-3 and 7-4) such differences to exist between the Moon and the Earth's crust – not only in the trace elements, but also in those of relatively high abundance – as to make it extremely unlikely that these were once a part of the same planetary mass.

In view of these facts, it becomes necessary to consider the only alternative open to us – namely, that the Earth and the Moon came into being simultaneously, as a part of the same creative process which gave rise to the entire solar system. Did these two bodies originate close together to be gravitational partners to each other since the time of their formation, or was the Moon captured by the Earth from a helio-centric orbit of its own; and if so, when?

Let us proceed to examine such possibilities in the light of all knowledge and information available to us at this time. The difference between both is, to be sure, largely semantic; for a simultaneous condensation in close proximity is tantamount to a capture if the relative velocity of both bodies was small. The possibility of a capture as such is not limited by such a constraint, and can be effected even from hyperbolic relative velocity – provided that a suitable sink of kinetic energy is on hand to reduce this velocity below the parabolic limit. The situation is quite analogous to the capture by the Moon of an artificial satellite sent out from the Earth, as discussed in Chapters 1 and 2. If such a satellite were to remain in free flight after its initial disengagement from the Earth, it would by-pass the Moon as a fly-by and revolve henceforward around the Sun. In order to enable the Moon to capture such a satellite in a closed orbit, retro-rockets have to be fired to reduce the kinetic energy of the fly-by.

In cosmic cases – such as represented by a pair of stellar or planetary bodies which happen to find themselves in close proximity of each other – the sink of energy necessary for a capture can be provided by other bodies in the proximity (confronting us with a many-body problem, such as we meet in star clusters); or – if the close pair is isolated in space – by *dissipative* forces arising from mutual *tidal interaction* between these bodies, which causes them to depart from spherical form.

For it is well known that if a self-gravitating body of finite size is subjected to attraction by an external mass situated at a finite distance, the force exerted by this mass will raise tides on the body subject to attraction, and vice versa. In order to outline the dynamical effect exerted by such tides on the relative motion of the pair – which may consist of the Earth and the Moon – suppose first that both these bodies possess perfect elasticity in their solid parts, or perfect fluidity if they were liquid. Under these conditions the equilibrium theory of tides reveals that the maximum height of the tides occurs in the direction of the attracting body, and also (for even-harmonic tides) in the opposite direction. Because of this symmetry, it can be shown that the lunar tides on the Earth (or the terrestrial tides on the Moon) would then perturb neither the rotation nor the motion of either body – save for the secular advance of the apsidal line of their relative elliptic orbit.

If, however, the bodies in question are not perfectly elastic (or, if fluid, of finite viscosity), the high tides raised mutually by them no longer point in the direction of the attracting body (or diametrally opposite to it), but lag behind this direction (or advance ahead of it) depending on the difference between the velocity of axial rotation and orbital revolution. Thus if tidal deformation of the Earth by the Moon is accompanied by friction (due to viscosity, or imperfect elasticity), the rotation of the Earth will carry the lagging tidal bulge forward (see Figure 10.1). Moreover, the gravitational attraction on this bulge is asymmetric with respect to the radius-vector joining the centres of the two bodies; this asymmetry gives rise to a torque tending to decelerate the Earth; and an equal and opposite torque tending to accelerate the Moon. Because of the friction this transfer is, moreover, accompanied by a loss of the mechanical energy of the Earth-Moon system (and its conversion into heat). In addition, as the Moon does not remain in the equatorial plane of the Earth, the tidal bulge is asymmetric with respect to the equator. As a result, one component of the torque acts constantly on the position of the orbital plane of the Moon; and an equal and opposite component tends to change the direction of the terrestrial axis of rotation.

Similarly, if the Moon's angular velocity of axial rotation were greater than the component of the Earth's rotation perpendicular to the orbital plane, the surface of the Moon would move ahead of the lagging tidal bulge, and angular momentum would be transferred from the Moon's orbital motion into the Earth's rotation. Should the component of the Earth's angular velocity perpendicular to the orbital plane equal the angular velocity of the Moon, frictional coupling between rotation

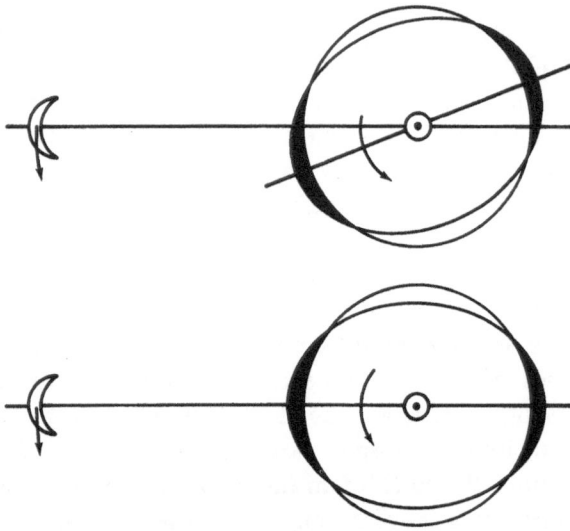

Fig. 10.1.   A schematic view of the tidal bulge raised by the Moon on the Earth under different conditions.

and revolution would disappear; though (in eccentric orbits) radial components of the tides would still continue to convert mechanical energy into heat at the expense of motion.

The most important characteristic of this tidal friction is, however, the fact that its effectiveness for transferring angular momentum between rotation and revolution depends so strongly on the distance between the attracting masses. In particular, when the Earth and the Moon are separated by the distance r, the height of the jth harmonic tides (of which that corresponding to $j=2$ is the most important) raised by one on the other is proportional to r^{-j-1} and, consequently, the corresponding tidal torque varies as $r^{-2(j+1)}$ for $j=2, 3, 4\dots$. Clearly, the secular changes in the orbital elements of the Earth-Moon system will be rapid when r is small, and become negligible as $r \to \infty$.

Is it possible that the Earth captured its satellite by such a process? In principle, yes; though any direct proof of this is still lacking. Can we, moreover, say anything about the time when this may have occurred? That tidal friction is indeed operative in the Earth-Moon system is supported empirically by the observed secular deceleration of the Earth's rotation and of the Moon's motion. According to a recent review of the relevant data by Munk and MacDonald (1960) the astronomical evidence points to a tidal lengthening of the day by 1.8×10^{-3} s per century, or a fractional change of 2.1×10^{-10} per annum. This figure has recently been tentatively verified in a remarkable way by Wells (1963), or Panella (1972) who found that certain ridges in the skeletal structure of corals might be evidence of a diurnal growth cycle. Recent corals show about 360 such ridges per year, while fossil corals from the Middle Devonian (age 400 million years) have about 400 ridges per year – corresponding to a fractional increase in the length of the day by 40 days in 4×10^8 yr, or 2.8×10^{-10} of a year per annum.

Accordingly, the semi-major axis a of the Moon's orbit should presently be increasing at a rate close to 3.2 cm yr; its inclination i diminishing by $1''.9 \times 10^{-6}$ per annum; and the eccentricity e increasing by 1.2×10^{-10} per annum. It is, of course, possible also to integrate the equations which govern the secular variation of the orbital elements backward in time, in order to find out what the orbit of the Moon around the Earth may have looked like in the more remote past – beyond the range of the 'book of the hours' established by the geologists. It is evident from the above estimates of the present orbital changes that, in more distant past, these changes could have been considerably faster – particularly if a becomes small (because the effects of the tidal torque varied as a^{-6} for the dominant second-harmonic tides, and even faster for higher harmonics).

The actual integrations of the respective variational equations, as performed in the pre-Apollo days by Gerstenkorn (1955, 1967, 1969), MacDonald (1964) or Singer (1968), were more of the nature of mathematical exercises; for they attempted to reconstruct the distant past of the Earth-Moon system on the assumption that the presently observed rate of tidal friction remained constant at all times. That this was so is unlikely; for the internal structure of the Moon, and even more so that of the Earth (whose relatively rapid axial rotation constitutes the main source of kinetic energy fed into the process) could have evolved appreciably in the course of the time; and the past rate of the tidal friction could have been quite different.*

On the assumption of constant rate of tidal friction extrapolable in the past, the above-mentioned investigators concluded that the closest approach of the Moon to the Earth should have taken place a little less than 3 billion years ago. We know today (cf. Chapter 8) that nothing particular has happened to the Moon at that time. If the Moon came ever as close to the Earth as it would have been necessary for raising tides sufficiently high for capture, so brutal an encounter should have left important markings also on the surfaces of the two bodies. It is, of course, futile to look for any such evidence on the Earth; for very little of its crust existing at that time has been preserved. But with the Moon it is very different; and yet no particular events of the history of its surface can be dated to that time.

The significance of this negative testimony is far-reaching. We know now that the hypothetical capture of the Moon by the Earth could have taken place only 4.6 billion

---

* For a summarizing discussion of this problem, cf. Kaula (1971); also Lyttleton (1967).

years ago or very close to it; and if we scale up the Gerstenkorn-MacDonald computations in such a way as to obtain the closest approach 4.6 billion years ago, we must *slow down* the presently observed rate of tidal friction in the past rather than to increase it. The only way to do so is to render both components of the Earth-Moon system – and, in particular, our mother planet – *less viscous* than it is today. Only the geophysicists can tell the astronomers whether such a hypothesis would be permissible; but if not, we shall be forced to conclude that the Moon was never close enough to the Earth to have been captured by it.

But while the decision is still in suspense, let us return to the lunar surface as the oldest surviving witness of the early days of the solar system, and delve further into the message which its stony face holds for us in store. We explained already in Chapter 8 that large parts of the lunar surface (i.e., its 'continents') solidified as far back as 4.6 billion years ago, and have not been re-melted since by any process. Moreover, as is well known, the stony relief of the lunar face represents a cumulative record of events – both external and internal – which disfigured the Moon since the time of its formation. As far as the continental areas on the Moon are concerned – and these constitute more than 80% of the entire lunar surface – the formations still extant were accreted probably in the first few hundred million years (possibly less) of the lunar existence; and the formation of the different maria on the Moon represented apparently a series of isolated episodes spread out between 3–4 billion years before our time; after which the morphology of the surface of our satellite seems to have undergone few if any important large-scale changes.

We now wish to examine, from this point of view, the testimony which the main features of the lunar surface – preserved in their pristine state by a virtual lack of any kind of erosion – bear on the cosmic environment in which our satellite acquired its principal morphological characteristics (cf. Kopal, 1972). In doing so we shall adhere to a working hypothesis that, in the earliest stage of the history of our satellite, the lunar sculpture in the continental areas was predominantly shaped by external impacts; and that, in particular, the majority of all large craters on the Moon represent surface scars produced at that time by primary intruders, then more abundant in space than they became (on account of depletion) in subsequent ages. Whether or not individual crater formations could also have been produced by other (internal) processes will be irrelevant for our argument as long as the numbers of such formations do not become statistically significant.

On the assumption, therefore, that the majority of old craters in the continental areas represent surface scars due to primary impacts, their morphological characteristics can tell something about the statistical distribution of directions from which our satellite suffered hits from outside. Extensive laboratory simulations of the cratering processes by Gault (1973) have disclosed that, for angles of impact in excess of (approximately) 15°, the characteristics of the astroblemes produced on the lunar surface can disclose next to nothing as to the direction from which the respective cosmic intruder approached our satellite. For oblique impacts inclined by less than 10°–15° to a plane tangent to the lunar surface the direction of the intruder can, how-

Fig. 10.2. A nearly vertical view (at high Sun) of the western part of Mare Foecunditatis, taken by the metric camera of Apollo 15 mission in July 1971 from an altitude of 120 km above the lunar surface. The two craters near the centre of the field are Messier and Messier A, associated with a twin bright ray akin to a cometary tail (a very appropriate accessory for Messier, who was a well-known comet hunter of his time) which is indicative of an oblique impact (NASA official photograph).

ever, be inferred from an elongated form of the resultant craters (see Figure 10.2), or from an unequal distribution ('zone of avoidance') of the apron of the ejecta scooped up by the impact (Figure 10.3).

Impacts oblique enough to preserve the characteristics of the original direction of

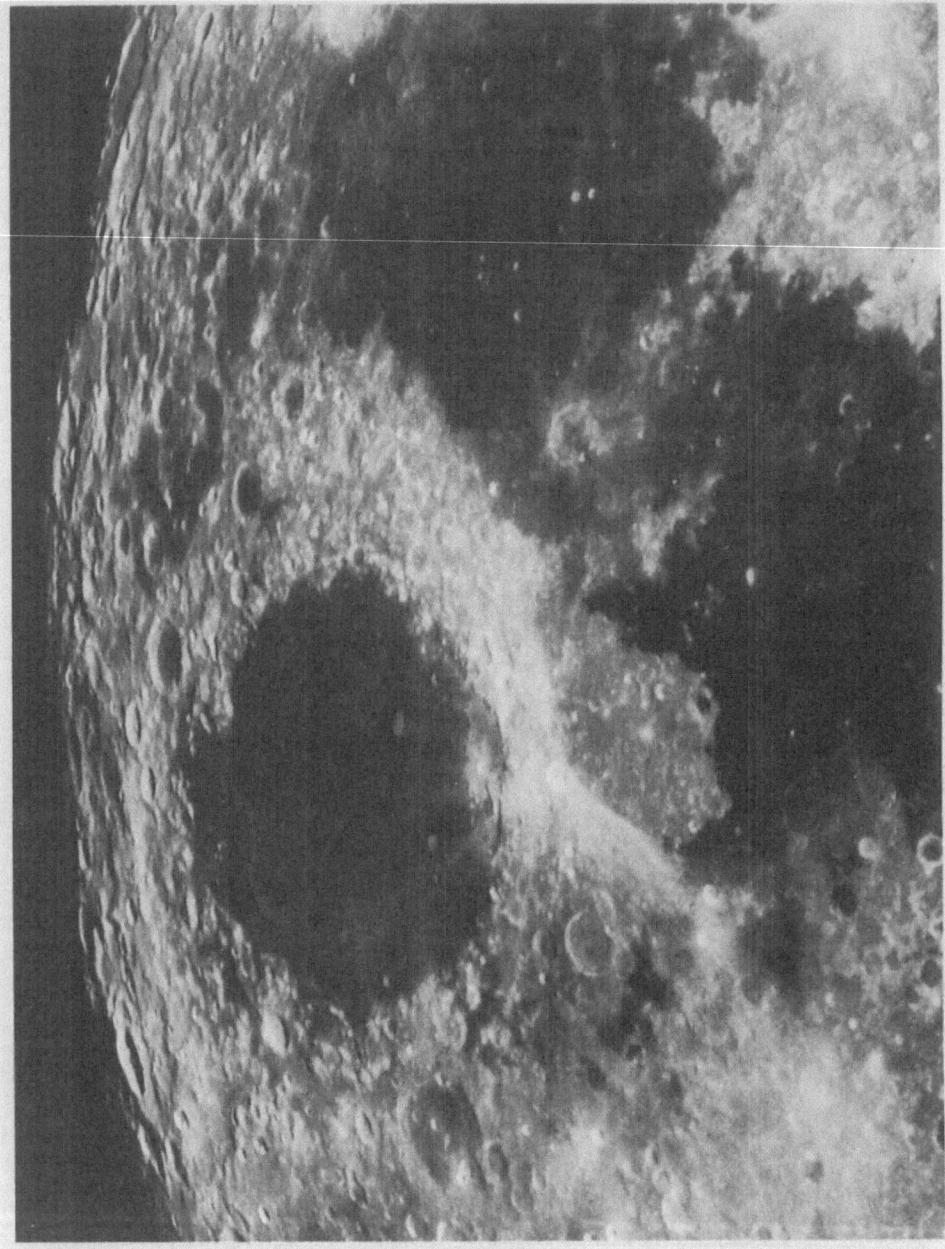

Fig. 10.3.   An asymmetric apron of bright ejecta from the crater Proclus on the western side of Mare Crisium (as photographed with the 24-in. refractor of the Observatoire du Pic-du-Midi), suggesting the effects of an oblique impact.

approach are, to be sure, likely to constitute only a very small fraction of all such events recorded on the lunar face. Nevertheless, the total number of impacts suffered by the Moon in the course of its long astronomical past from all directions is so large that the number of oblique events among them should not be negligible. In point of fact, directional characteristics of low-angle impacts are indicated for a few hundred crater formations exceeding 1 km in size (a few dozen of which are larger than 10 km) on the front side of the Moon alone; and those of this type smaller than 1 km are too many to be individually counted.

Such being the size of the sample available for statistical analysis, the question is bound to arise: what is the distribution of low-angle impacts which the Moon has suffered since the time of its formation? The stony sculpture of the continental areas goes back, in its structure, probably all the way to the age of 4.6 billion years established for the bulk of the lunar 'fines' returned by successive Apollo missions of 1969–1972; and while large-scale features (in excess of, say, 100 km in size) may have partly fallen prey to the 'saturation bombardment' suffered at the earliest stage of lunar history (cf. Marcus, 1966), this is not likely to be true of the smaller crater formations preserving directional characteristics. Therefore, their cumulative sample available for our analysis should have reached us essentially unimpaired from the earliest days of lunar history, and provide a link with the most distant past of the solar system.

The answer to our question turns out to be simple, but very interesting in its cosmogonic implications: namely, *the distribution of craters due to oblique impacts appears to be essentially at random all over the lunar surface* – and does not exhibit clustering in any particular zone. The full significance of this result will emerge only when we consider it in connection with the orientation of the Moon in space, and the direction of its axis of rotation with respect to the invariable plane of the solar system.

As is well known, the inclination I of the lunar equator to the ecliptic is very small – only $1°32'4'' \pm 7''$ according to its latest determination by Koziel (1967a, b). An even more meaningful result will be obtained if we refer this inclination to the invariable plane of the solar system, the present orientation of which with respect to the ecliptic is specified by the angles $\Omega = 106°44'$ and $i = 1°38'58''$ for the epoch of 1900.0 (cf. Brouwer and Clemence, 1961). Koziel's value of I given above, and deduced from the lunar heliometric observations carried out between 1877–1915, should refer approximately to the same epoch; and if so, the inclination of the Moon's axis of rotation (as specified by the angles $I = 1°32'4''$ and $\Omega = \pi$) to the invariable plane of the solar system around 1900.00 must have been equal to $I_{1900} = 2°1'$.

This latter value cannot, to be sure, remain secularly constant; for the longitude of the node of the invariable plane is known to be increasing (at present) by $59'$ per century, and its inclination diminishing by $18''$ per century (cf. Brouwer and Clemence, 1961) – no doubt as a result of an action of harmonic terms of very long period – and while nothing as yet is known directly about the nutation of the lunar axis of rotation, its amplitude is unlikely to be zero, and its period must be very small a fraction of the total age of our satellite. Therefore, all we can say at present about the inclination of the lunar axis of rotation to the invariable plane of the solar system is that it oscillates between $i \pm I$, or $0°-3°$; and may have done so throughout the entire astronomical past of our satellite. But if so, it would follow that its equatorial belt, or polar caps, have never been very far from the positions which they occupy at the present time; and their latitudes may have secularly changed by only a few degrees.

Now let us examine, in this light, the fact that evidence for impacts in all directions – including low-angle or grazing impacts – appears to be distributed at random all over the surface of our satellite; and inquire about the orbits which the impacting particles must have followed prior to their collisions with the Moon. If these particles had been in heliocentric orbits, the inclinations of such orbits could not have deviated much from the ecliptic, or they would have missed the Moon altogether; for only particles in orbits closely parallel with the ecliptic could have impinged anywhere on the surface of our satellite. But if the Moon's axis of rotation has been almost perpendicular to the ecliptic, it would follow that *high-angle impacts should have occurred primarily in the equatorial zones; and low-angle or grazing impacts limited almost exclusively to the polar regions of our satellite.*

The extant stony record of the lunar face discloses conclusively that this is *not* the case; but, rather, that impacts at all angles appear to be distributed at random in all parts of the Moon. But if so, it follows that *the impinging particles could not have revolved in heliocentric orbits,* but must have been moving in trajectories which made impacts at all angles equally likely all over the surface of our satellite. In other words, the particles destined to end up their cosmic careers by collisions with the lunar surface were not moving prior to impacts around the Sun, but within the domain of influence of the Earth-Moon gravitational dipole; with the Sun exerting only perturbations of their motions at a distance. Only in this way could impacts in all directions become equally likely.

This – if true – would suggest that the particles responsible for the bulk of the cosmic bombardment of the lunar surface at the earliest stage of its history were those left over in the Earth-Moon neighborhood from the process which led to the formation of this pair of cosmic bodies, and carried along with them through space until most of them were swept up by the gravitational attraction of the two mass centres. This would, in turn, imply that the Earth and the Moon were already gravitational partners (though not necessarily at their present distance) during the first few hundred million years of their existence – an epoch in which the cratering of the lunar surface is now thought to have been largely accomplished – and if the Moon was captured by the Earth, this event must have occurred prior to the main period of cratering of the lunar continental land masses.

Is there any escape from these conclusions? One would be to assume that most part of the continental impact craters were inflicted on the Moon before its eventual capture by the Earth – when the Moon, travelling alone through space, may have been tumbling about its centre of mass, so that impacts from all directions would have been equally likely.

Such a view is, however, highly improbable. It is true that the present rate of axial rotation of the Moon, and its physical librations in latitude, longitude, and node are virtually controlled by the Earth (and would have been so even more effectively at a shorter distance between the two bodies). We have, therefore, no idea whether or not the Moon could have been spin-stabilized in space before its hypothetical capture. However, even if this was not the case, it is most unlikely that any of its pre-existing surface sculpture would have survived so brutal an experience as a capture of the Moon by the Earth would inevitably have been. The need to slow down a passer-by to make capture possible would have required a dissipation of so much energy of the Moon through inelastic tides as could be raised only at a very close approach; and such tides would no doubt have obliterated most, if not all, of the pre-existing surface markings.

Therefore, if the Moon was captured by the Earth at the earliest stage of the history of the solar system, its present surface sculpture must have begun to accumulate after the time of the capture; and if it was formed as an independent body in the proximity of the Earth to begin with, its inclination to the invariable plane of the solar system has probably remained more or less unchanged up to the present time. But, either way, the argument against heliocentric orbits of particles whose collisions with the Moon disfigured most part of its continental areas continues to hold good.

It should be stressed that our Moon represents the only planetary body of our system, within easy observational reach, to which this argument can be applied. The space missions to Mars, commencing with the historical flight of U.S. Mariner 4 in 1965, have disclosed the surface of the sister planet of ours to be pock-marked with craters of similar kind – and, no doubt, similar origin – as those which have disfigured our satellite. However, unlike for our Moon, the axis of rotation of the Martian globe is considerably more inclined to its orbital plane – the inclination of the Martian equator to the ecliptic now amounts to 23°51' (in contrast with 1°32' for the Moon), and may likewise oscillate somewhat on account of nutation. Moreover, the oblateness of the Martian globe is bound to give rise to a precession of the Martian axis of rotation, due to solar attraction, with a period of the order of 10^6 yr which – long as it is in comparison with that of the luni-solar precession of our own planet – is fleetingly short in comparison with the age of the solar system.

Therefore, in the course of its long astronomical past, the directionality of impacts on the Martian surface would have been largely 'de-focussed' by the precession of its rotational axis in space. There is but little room for doubt that cosmic particles, whose collisions with the Martian surface gave rise to such impact craters as we observe there today, were in heliocentric orbits prior to their fatal encounters with our sister planet. As, however, the direction normal to any element of the Martian surface may oscillate by $\pm 24°$ in the course of each precessional cycle, the directionality effects of particles in heliocentric orbits could be very largely smoothed out in the course of time. It is the near-perpendicularity of the lunar axis of rotation to the ecliptic which endows our Moon with particular significance in this respect, and renders its present testimony so valuable for the dating of the commencement of its gravitational association with the Earth. Summarizing it, we conclude that the Moon is no true-born child, but most probably a foundling of our planet – captured and adopted as such before the main features of its present stony sculpture was impressed over its continents; and this sets the time of the commencement of this long-lasting liaison to no more than a few dozen million years after the origin of this solar system.

To those astronomers who may consider such a capture an unlikely event, we should point out to our sister-planet Mars which possesses, not one, but two even more unlikely satellites (Phobos and Deimos) which cannot be anything else but captured asteroids. How could, however, two such unlikely satellites have been captured? Tidal friction may possibly be effective enough to this end with a satellite as massive as our Moon; but manifestly not for such tiny freaks as Phobos or Deimos. The only reasonable suggestion as to the circumstances of possible capture was advanced by Urey, who proposed that two asteroids from their nearby belt happened to pass close to Mars – or, rather, proto-Mars – before the completion of its formation as a planet we know today; and the dissipation of kinetic energy needed for capture could have been accomplished inside the proto-cloud.

For Mars this seems – we repeat – to be the only reasonable proposal one can put forward at the present time; but whether or not a similar process could also be invoked to account for the Moon's capture by the Earth remains uncertain; our Moon may be too massive for such an operation. Besides, the Earth need not have captured our only natural satellite at all – and the latter could have coalesced from a circumterrestrial swarm in orbit around the Earth at the same time as the central planet was formed (cf. Ringwood, 1970). According to this view – which is at present being developed by the Russian school of investigators (cf., e.g., Ruskol, 1960, 1963, 1966, 1971, 1973) – the original agglomeration of mass of a proto-planet which eventually gave rise to the Earth-Moon system failed initially to collapse all towards its centre of mass; and a part of it remained outside to form a ring revolving around the centre which later coalesced into the Moon. But whatever the case may be, there seems no escape from the conclusion that the Earth and the Moon must have been gravitationally coupled together for almost their entire astronomical past, and did not form their liaison at much later date.

2. Origin of the Maria

According to the view just expressed, the predominant part of the stony sculpture of the Moon's continental land masses cannot be far removed in age from 4.6 billion years – most of it probably originated no later than 4.6–4.3 billion years before our time by multiple impacts of debris left about from the days of the formation of the system (a supply of which must gradually have been diminishing). Such a picture would, however, not yet include some of the largest and most spectacular impact events which the Moon experienced in its long history: namely, those which produced the large basins on the lunar surface, some of which became the lunar maria.

A view seeking to explain the origin, not only of lunar craters, but also of the maria by external impacts was originally proposed by Gilbert (1893), and subsequently championed by Urey (1952, and many subsequent writings*) – in particular, with respect to the Imbrian basin – so comprehensively that he should be regarded as the real spiritual father of this concept. It will be the aim of this section to survey critically the present state of this concept, and to assess the extent to which it can provide the right clue to subsequent epochs of lunar history.

Before the advent of the space age, lunar maria – in particular, those of circular form – were regarded as being nothing else but very large impact craters which differ from smaller formations of this type (such as the craters Copernicus, Theophilus, or Tycho) in size rather than in kind. The mountain chains bordering their shores (such as the lunar Apennines, Alps, Juras, and Carpathians encircling Mare Imbrium) were then considered to constitute ramparts of the respective formation – broken into isolated ranges for large maria like Mare Imbrium or Serenitatis, or forming a continuous ring for smaller maria like Mare Crisium or Moscoviense.

When the empirical relations between the linear size of a crater and the energy necessary for its production have been extrapolated to formations of mare size, the kinetic energy of bodies required to produce them on impact could be estimated. For Mare Imbrium – almost 700 km across – this energy turns out to be of the order of 10^{32} erg; and if the relative velocity of the impinging body did not exceed much that of escape from the gravitational field of the Moon (i.e., 2.4 km s^{-1}), the mass of the intruder should have been almost 10^{22} g (i.e., about 10^{-4} of that of the Moon; corresponding to a good-sized asteroid). Moreover, the amount of heat which would have been produced by a sudden stoppage of such a projectile would have been sufficient to melt an amount of material sufficient to cover the whole expanse of Mare Imbrium with a layer of molten lava a few hundred metres in depth (see, e.g., Arkani Hamed, 1974).

Since the advent of the space age a number of new facts have, however, come to light which required substantial revisions of this simplified picture. Thus when the 1959 Luna 3 photographs unveiled for us a first glimpse of the Moon's far side, it transpired that the dark-filled maria are conspicuous by their absence there, and seem concentrated predominantly on the visible hemisphere of our satellite. Moreover, from the Zond 3 photographs of 1965 we learned first (cf. Lipsky, 1965) that basins exist on the far side of the Moon, of the dimensions of lunar maria, which are not filled with any lava (the Russians have called them 'thalassoids'). Since that time, several additional basins of this kind have been discovered on both sides of the Moon from subsequent space-borne photography.

The discovery, in 1968, of distinct mass condensations at relatively shallow depths below the floors of the circular maria strengthened the case for impact origin of such formations and of lunar basins in general; but at the same time gave rise to other questions which are not yet easy to answer. For example, if the dark filler covering the floors of lunar maria were to represent a lava melted by the impacts, why should

---

* For the latest statement of his views, cf. Urey and MacDonald (1971).

only certain maria (predominantly on the front side of our satellite) appear to have been flooded while others of comparable magnitude were left in a 'dry' state?

And, secondly, why do the lunar basins (including the maria) not depart much more conspicuously from the spherical shape of the lunar globe? If a basin of the size of Mare Imbrium – 700 km across – had been excavated by an impact differing only in scale from those which produced smaller craters, so large a basin should have been also more than 200 km deep (i.e., a depth well below the 80-km discontinuity in the structure of the lunar globe, indicated by the seismic evidence already discussed in Chapter 5, and below the depth of the associated mascon). The surface of the Mare Imbrium (see the altimetre profiles shown on Figures 4.1 and 4.2) does not deviate from a mean sphere by more than a few kilometres; and (as we shall demonstrate later) its lava fill is not more than a few hundred metres deep. What, therefore, could have uplifted the floor of the initial precipice by 200 km? Isostatic adjustment is difficult to think of in view of the rigidity of the lunar crust necessary to support the mascons and other deviations from hydrostatic equilibrium discussed in Chapter 4; but if not that, what else?

Returning to the mascons – since 1968 it has been gradually realized that the first views regarding them as mere leftover masses of the original intruders were probably ill-founded. For once, the mascon masses turned out to be about 10 times smaller than those required to scoop up by impact a basin of the respective mare size (cf. Urey and MacDonald, 1971). Besides, lunar gravimetric work of recent years indicated that smaller formations of impact origin (such as 100 km craters of the calibre of Copernicus or Theophilus) exhibit *negative* gravitational anomalies – in contrast with the positive anomalies characteristic of the circular maria. Therefore, a difference in size between these formations seems to indicate also a different structure in depth; but the reasons for this difference are as yet unclear.

The structure and age determinations of the lunar rocks imported from the mare sites by the Apollo and Luna missions between 1969–1972 threw further important light on the origin of lunar maria, and helped to clarify some of the problems mentioned in the earlier paragraphs; but, at the same time, confronted us also with new problems which will be pointed out below.

On one hand, a large preponderance of brecciated rocks at these sites left but little room for doubt that the lunar basins are indeed of external origin, and due to impacts which occurred intermittently between 3.9–4.2 billion years before our time. On the other hand, the mare basalts which subsequently filled up several of these basins (not all – especially on the Moon's far side) turned out to be several hundred million years younger than the breccias. If, therefore, the age of the breccias determines the time of the impact event, it was clearly not the impact heat which melted the magma now covering mare floors; for the latter must have exuded from the interior at a considerably later time. Yet the heat generated by impacts must have played a role of its own in shaping up the internal structure of the respective regions. Was it this heat which facilitated a partial return to hydrostatic equilibrium that lessened the original depth of the 'dry' basins?

A quest for the source of the basaltic magma we now see spread over the surface of the mare basins should, therefore, lead us back again into the lunar interior, which we left in Chapter 5, in quest of subterranean lava pockets which should have existed in the Moon in the fourth billenium of lunar history. It should be stressed that their existence is in no way at variance with the dynamical arguments set forth earlier in this chapter against a global ocean of magma covering the entire Moon; localized pockets are certainly immune to them. Computations of the thermal history of the Moon, carried out in recent years e.g. by Toksöz *et al.* (1972, 1973) or Duba and Ringwood (1973), have shown that such pockets can indeed originate under the right assumptions.

On the other hand, computations of the thermal history of the lunar globe have so far passed but little the stage of numerical exercises; for not enough constraints are known from the observations to render the problem determinate; and particular features of the solutions obtained depend largely on our underlying assumptions. This is true of the assumed chemical composition of the Moon at the time of its origin (which does not appear to have been either solar, or chondritic) and, in particular, of the abundance of different radioactive elements in its material. Nevertheless, recent computations by Ringwood, Toksöz, and others have shown that, under assumptions which are not implausible, sub-surface melting could have occurred on the Moon at the right time; and, to this extent, theory does not oppose the view that lunar maria are covered with lava flows of internal origin.

Greater difficulties are encountered, however, when we consider the mechanism which could have driven out such lavas to the surface. The petrographic structure of the lunar mare basalts is such as to indicate that they were differentiated under pressures close to 10 kb, obtaining inside the Moon at a depth of not less than 150 km i.e., larger than the depth of volcanic chambers of the terrestrial volcanoes. To pump molten material from such depths up to the surface calls for an expenditure of energy whose source is not easily apparent.

This task may be facilitated on the Moon by a relatively low gravity, and also by the fact that, as far as we know, molten magma flooded vast plains of the lunar maria down to only very shallow depth – not exceeding in general a few hundred metres. That this is so we know from the dimensions of 'ghost craters' or other formations pre-existing on the former floor of the respective mare, whose ramparts after flooding still protrude above the level of the present mare. Such ghosts abound in many parts of the lunar maria, and an example of one south of the crater Lambert in Mare Imbrium, is shown on Figure 10.4. The diameter of this ghost is easily measurable when the Sun stands low above the horizon; and since, for most craters, their diameter bears a certain ratio to the height of the ramparts ('Ebert's rule'), it is possible to estimate from the measured diameter the depth at which the respective crater has been submerged.

A determination of such depths from a number of ghost craters whose features lend themselves for this purpose disclosed (cf., e.g., De Hon, 1974) that basaltic lavas cover most lunar maria only to the depth of a few hundred metres. Thus the average thick-

Fig. 10.4.   The photograph of a ghost crater (Lambert R) south of the crater Lambert, taken by the metric camera of the Apollo 15 mission in July 1971, from an altitude of 98 km.

ness of the mare fill in Mare Tranquillitatis appears to be between 500–600 m; with maximum accumulation no more than twice this amount. Mare Imbrium does not appear to be any deeper; only Mare Nectaris attains an average depth of some 900 m, and maximum depth close to 1500 m.

The fact that the basaltic lavas seem to be spread out over extensive mare plains in the form of rather thin veneer makes it possible better to understand certain other features of their topography disclosed by laser altimetry. It goes without saying that, if the basaltic magma which filled the pre-existing basins of the present maria had been molten at any one time, it should have solidified in the form of a level surface which is everywhere normal to the local gravity.

The actual vertical profiles of the ground tracks overflown by the Apollo 15–17 laser altimeters disclose, however, that this is not quite the case. In point of fact (cf. Figure 4.1), the present orientation of several mare plains – in particular, of Maria Serenitatis and Imbrium – appear to be noticeably inclined to the horizontal direction; their inclinations being parallel, and such as to render the western shores of the Imbrium-Procellarum complex to lie about one kilometre higher than the eastern shores. If these were deep pools of lava that solidified at any one time, such a situation would be incomprehensible; and can be understood only if the form of the present surface is governed by the slope of its solid substrate rather than by gravity (as it would be for the boundary of freely flowing liquid).

However, the laser altimeter output of Apollo 15–17 confronted us also with other facts which raise important questions as to the mobility of molten magma over marial surfaces. In order to demonstrate these, attention is invited to the accompanying Figure 10.5, on which we have reproduced a photograph of the western parts of Mare Imbrium, as recorded by the 'metric' camera of Apollo 15 from the orbital altitude of 112 km. With the Sun very low above the horizon, this photograph shows beautiful examples of what have been generally regarded as petrified magma flows, arrested in their motion by solidification on cooling. It goes without saying that, should this be the case, the direction of such flows would have to be normal to the local isoclines – since molten magma (like any other liquid) can flow only downstream, and not uphill.

The present writer pointed out on previous occasions (cf., e.g., *Moon II*, pp. 466–467) that evidence for this was rather dubious; and more recent Apollo hypsometric data have vindicated these doubts. In particular, it appears now that the so-called 'lava flow' formations shown on Figure 10.5 are not directed downstream, but approximately along the local isohypses; and if so, their present orientation can scarcely represent that of an arrested lava flow. The writer considers it more likely that such formations represent essentially static features, and are due more to vertical warping of the surface at the time of its solidification than to any horizontal flows. In other words, the hypothetical lava flows may actually be 'blisters' formed by non-uniform cooling of the area during its transition from liquid to solid state.

Other problems arise, furthermore, in connection with the Apollo seismic data which becloud further the issues: namely, how to account for the present fractured structure of the floors of the lunar maria. In Chapter 5 we already presented the evidence – mainly in the form of the anomalously long durations of lunar seismic echoes – that the first 20 km of the lunar crust (in particular, in the maria where three Apollo seismic stations have been installed) must consist of layers which are not solid, but highly fractured to provide a broken medium in which seismic waves can undergo the requisite amount of scattering.

What mechanism could have fractured the underlying mare floors to such an extent? The radioactive dating of the rocks imported from such localities disclosed that

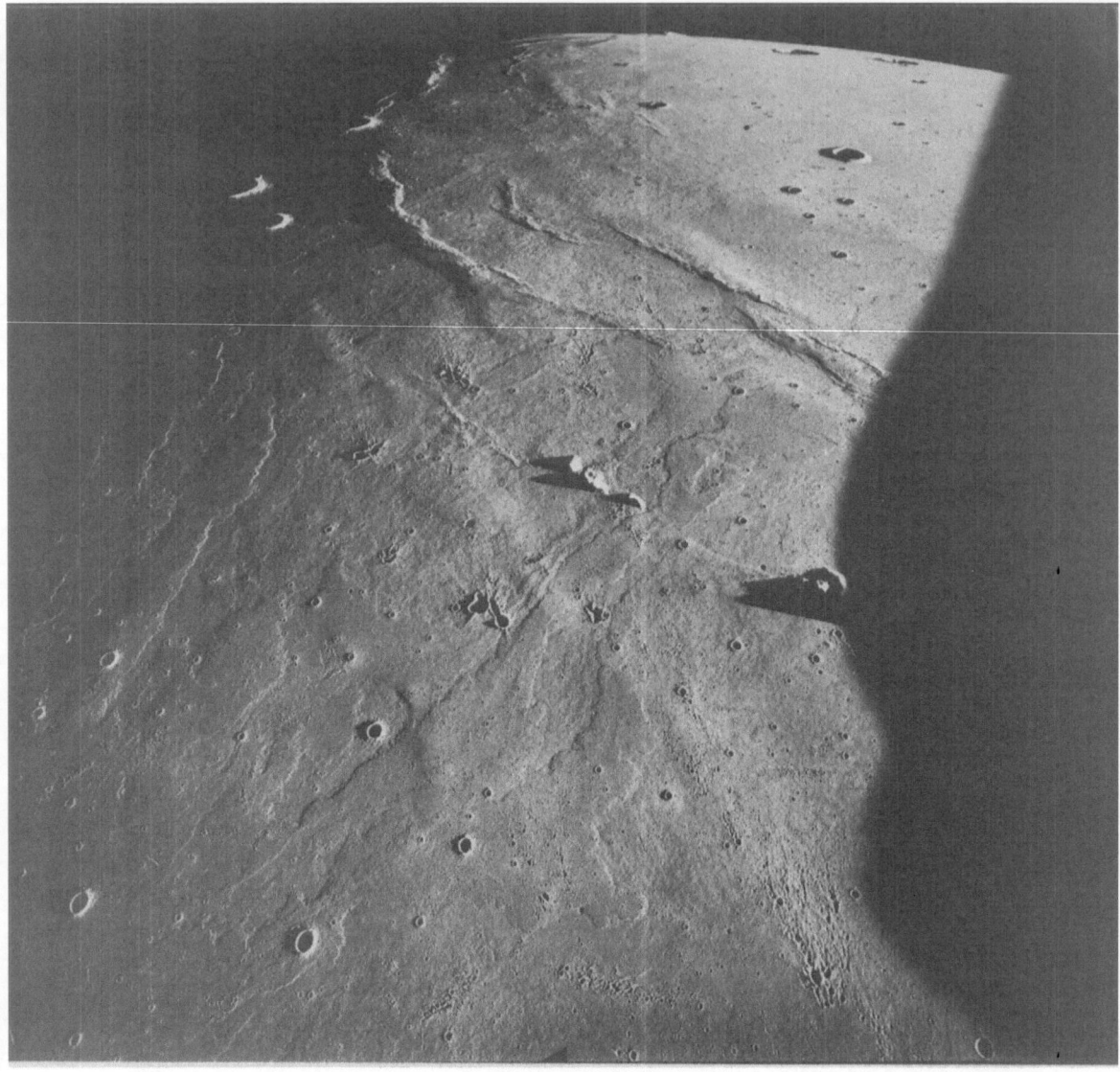

Fig. 10.5. A photograph of the eastern plains of Mare Imbrium (near the crater Herschel), taken by the metric camera of the Apollo 15 mission in July 1971 from an altitude of 99 km above the lunar surface (NASA-Houston photograph).

they solidified (at least, for the last time) at discrete times given in Table 8-2 – which, for Mare Imbrium, appears to be close to 3.3 billion years; and 3.4 billion years for Mare Foecunditatis. Whether or not the chemical differentiation manifest in these samples occurred at that time, or whether the material then remelted was already differentiated, is besides the point. Essential is the fact that the cooling which followed this latest melting should have led to the formation of a solid mare floor at that time.

Seismic evidence now in our hands demonstrates, however, that mare floors are no longer solid, but highly fractured down to considerable depth; and the question arises as to the process which could have ended up in this way. Should we wish to account for the present fractured state of the lunar crust (including that of the maria) in the cumulative effect of external impacts by asteroids, meteorites, or comets subsequent to the last solidification of the respective mare floor we run, however, into two kinds of difficulty. First, the solidified mare ground does not exhibit sufficiently high crater counts which could account for the fractured nature of the underlying mare ground. Secondly, no evidence has been found for similar fracturing of the terrestrial rock strata of comparable age.

The data listed on Figure 8.7 make it evident that, by the time of the formation of at least some of the lunar maria, the geological history of the Earth and the Moon already began to overlap; and rock strata are known on the Earth (in Greenland, Canada, or Africa) which are fully as old as Mare Imbrium or Tranquillitatis. But these terrestrial strata show no evidence of extensive fractionation or cratering. Smaller mechanical damage could have been healed in them by the combined effects of the terrestrial air and water; but for larger impacts this explanation is insufficient. We believe that by the time when these terrestrial strata solidified, the cosmic bombardment of planetary surfaces was largely over; as the supply of missiles in interplanetary space was gradually running out. But if so, which events could have fractured the lunar surface down to a depth of some 20 km, disclosed by the seismic data, since that time?

One solution – proposed recently by Dainty et al. (1974) – assumes that the fractured rock layer, which produces most part of the observed scattering of lunar seismic waves, is overlaid in the maria by only a thin lava veneer which behaves like an elastic medium, but on account of its shallow depth acts only as a 'thick boundary' of the underlying fractured layer which produces the scattering; and while the origin of its fractionation may indeed be ancient, more recent lava flows need no longer be fractured to the same extent. Such a hypothesis would appear to be in line with other evidence already discussed (inclined levels of some maria !) indicating that the marial lava cover is indeed relatively thin – a few hundred metres, in contrast with the mean depth of some 20 km estimated for the scattering layer. This may indeed represent a way out of this particular dilemma; but whether it constitutes its correct solution must await the verdict of further independent evidence.

3. The End of the Road

The fact that we are not yet in a position to answer satisfactorily any of these questions shows the distance we still have to travel before even the main problems of the past evolution of our satellite can be regarded as well understood. Fortunately, having outlined its probable past history for the first 1.3 billion years of its existence, we are approaching also the end of our task; for by the end of the first quarter of the Moon's existence, the active part of its history was just about over; and nothing more of major importance seems to have occurred on the lunar surface since that time.

To be sure, the Moon continues to expose its wrinkled stony face – weathered by age – to all cosmic influences which may act upon it; and keeps accumulating scars of new impact craters, but at a diminishing rate; for after 4.6 billion years of continuous gravitational sweeping of the interplanetary space by all planets, not many left-overs are around with which the Moon can be on a collision course. Of its more conspicuous impact craters, only Copernicus, Theophilus, and Tycho originated probably in the past billion years; and it is quite possible that no more craters of this size will be added to the lunar pantheon in the future to mark the burial place of another small asteroid.

Therefore, in the aeons of time to come, the face of the Moon will become more and more withdrawn in its petrified grimace, reflecting the state of the solar system in the days long gone by – a true fossil reminder of the distant past. Moreover, tidal friction operative in the Earth-Moon system will continue to increase the size of its orbit by a gradual transfer of the rotational momenta of its constituents to their orbital momentum.

The axial rotation of the Moon is already now virtually synchronized with its revolution around the Earth (and is, in fact, bound to have been at all times; cf. Kopal, 1972), so that the only dissipation of energy which can occur inside the Moon is through the effects, on bodily tides, of the orbital eccentricity of the relative lunar orbit. On the other hand, our Earth is (on account of its much larger mass) still far from this state – rotating as it does 27.3 times faster than the Moon revolves; and the relatively large angular momentum of its axial rotation continues to represent a fair prey for the slow but relentless dissipative action of lunar tides.

The tides – both bodily and oceanic – raised on the Earth by the attraction of our satellite tend inadvertently to increase the length of the day by a gradual transfer of its rotational momentum to the orbital momentum through frictional forces – and will continue to do so until, at long last, a state of synchronism between rotation and revolution will have been attained not only for the Moon, but also for the Earth. At the end of this final act of tidal evolution, the relative orbit of both bodies will become circular, and approximately 1.59 times larger than it is at present. The tides raised by each body on the other will then become of the equilibrium type – stationary with respect to the surface – producing no torque; and in the absence of any outside dissipative forces the Earth-Moon orbit should become immune to any further change.

When this time has been reached, the duration of the day on both the Earth and the Moon will become equal to their orbital period of close to 55 days (or two months) of our present time; and the axes of rotation of both components will have been rendered perpendicular to the orbital plane. Under these conditions, the Moon will appear to us much smaller in the sky than it does today (its angular diameter should be only 19.'5, in contrast with its present mean value of 31.'1), and will continue to show the Earth exactly the same immutable face – as the Earth will to the Moon.

Not perhaps a particularly exciting end of a long romance which must have commenced with a much closer gravitational embrace 4.6 billion years ago, and which (some would maintain) created much heat in both partners at that time. But such seems to be the end of most long-lasting liaisons – in the sky as well as on the Earth.

The only grace which may save us from its sheet monotony would be if perturbations by the Sun, expanding in its post-Main Sequence stage, can eventually bring about a formal dissolution of the union consistent with the laws of celestial mechanics. But it may take still 2–3 times the present age of the system before this stage of eventual expiation for the original capture may be attained; and no one of us will – I am afraid – be around by that time to see it.

And in concluding this book on this sobering tone, the author would like to address the following words to future investigators and future students of lunar science. In whichever part of the vineyard you may be labouring to advance our common cause, be wary of partial solutions which may satisfy a particular set of constraints imposed by any single scientific discipline. Never lose sight of the fact that a grand solution of our problems can emerge only from a consideration of all aspects of the available evidence, and not only of those with which you happen to be most familiar.

The past course of our science is indeed replete with examples of pitfalls of which one can become a victim by ignoring our advice; and several were mentioned earlier in the text. The classical example is represented by the efforts of the past generations of geologists to regard the Moon merely as a miniature Earth, and to identify on it – mainly by analogy – various formations which are familiar to us on our terrestrial abode. It was not till when the Moon has been considered in its wider cosmic setting, and exposed to inevitable interaction with the full range of particulate contents of interplanetary space, that a history of its surface has gradually begun to emerge in a more realistic form.

But again to consider the Moon as a mere inhabitant of the solar system may not yet be enough. In the course of its long past of approximately 4.6 billion years, the Moon has more than twenty times accompanied our Sun on its journey around the centre of our Galaxy, and repeatedly experienced also a variety of galactic climates which may have left distinct landmarks – both macroscopic and microscopic – on its surface. On account of its age, it is clearly insufficient to consider the Moon exclusively as a denizen of our solar system, without considering at the same time interstellar influences to which it was likewise bound to be exposed – the dimensions of the solar system are certainly small in comparison with the mean free path of a whole spectrum of the particulate contents of the interstellar space that can interact with the lunar surface no less decisively than interplanetary matter.

And last but not least, it hardly needs to be re-iterated that the origin of the Moon and its present cosmic station is inextricably connected with the origin of the solar system as a whole, and cannot be separated from it. As a result, it would be quite unrealistic to expect that a full understanding of many problems mentioned in this last chapter of our book will emerge separately from the rest of a story of the creation of the entire planetary system, in which the origin of the Moon can constitute but an episode – it is only the amount of information we received from the Moon in the past few years which renders this particular episode of such superlative interest at the present time.

References

Arkani Hamed, J.: 1974, *Moon* **9**, 183.

Brouwer, D. and Clemence, G. M.: 1961, *Methods of Celestial Mechanics*, Academic Press, London and New York.

Dainty, A. M., Toksöz, N., Anderson, K. R., Pines, P. J., Nakamura, Y., and Latham, G.: 1974, *Moon* **9**, 11.

De Hon, R. A.: 1974, in *Lunar Science* V, Pergamon Press, Inc., in press.

Duba, A. and Ringwood, A. E.: 1973, *Moon* **7**, 356.

Gault, D. E.: 1973, *Moon* **6**, 32.

Gerstenkorn, H.: 1955, *Z. Astrophys.* **36**, 245.

Gerstenkorn, H.: 1967, *Icarus* **7**, 160.

Gerstenkorn, H.: 1969, *Icarus* **11**, 189.

Gilbert, G. K.: 1893, *Bull. Phil. Soc. Wash.* **12**, 241.

Hagihara, Y.: 1972, *Celestial Mechanics*, Vol. 2, MIT Press, Cambridge, Mass.

Kaula, W. M.: 1971, *Rev. Geophys. Space Phys.* **9**, 217.

Kopal, Z.: 1972, *Moon* **5**, 200.

Koziel, K.: 1967a, *Icarus* **7**, 1.

Koziel, K.: 1967b, in Z. Kopal and C. L. Goudas (eds.), *Measure of the Moon*, D. Reidel Publ. Co., Dordrecht, pp. 3–11.

Lipsky, Yu. N.: 1965, *Sky Telescope* **30**, 338.

Lyttleton, R. A.: 1967, *Proc. Roy. Soc.* A **296**, 285.

MacDonald, G. J. F.: 1964, *Rev. Geophys.* **2**, 467.

Marcus, A. H.: 1966, *Icarus* **5**, 165, 178, 190, 590.

Munk, W. and MacDonald, G. J. F.: 1960, *Rotation of the Earth*, Cambridge Univ. Press, Chapter II.

O'Keefe, J. A.: 1969, *J. Geophys. Res.* **74**, 1969.

O'Keefe, J. A.: 1970, *J. Geophys. Res.* **75**, 6565.

Panella, G.: 1972, *Astrophys. Space Sci.* **16**, 212.

Ringwood, A. E.: 1970, *Earth Planetary Sci. Letters* **8**, 131.

Ruskol, E. L.: 1960, *Astron. Zhurn.* **37**, 690.

Ruskol, E. L.: 1963, *Astron. Zhurn.* **40**, 288.

Ruskol, E. L.: 1966, *Icarus* **5**, 221.

Ruskol, E. L.: 1971, *Astron. Zhurn.* **48**, 819.

Ruskol, E. L.: 1973, *Moon* **6**, 190.

Singer, S. F.: 1968, *Geophys. J. Roy. Astron. Soc.* **15**, 205.

Sonett, C. P. and Mihalov, J. D.: 1972, *J. Geophys. Res.* **77**, 588.

Toksöz, N. and Solomon, S. C.: 1973, *Moon* **7**, 251.

Toksöz, N., Solomon, S. C., Minear, J. W., and Johnston, D. H.: 1972, *Moon* **4**, 190.

Urey, H. C.: 1952, *The Planets*, Yale Univ. Press, New Haven.

Urey, H. C. and MacDonald, G. J. F.: 1971, in Z. Kopal (ed.), *Physics and Astronomy of the Moon*, Academic Press, New York and London, 2nd Ed., pp. 213–289.

Wells, J. W.: 1963, *Nature* **197**, 948.

Wise, D. U.: 1963, *J. Geophys. Res.* **68**, 1547.

Wise, D. U.: 1969, *J. Geophys. Res.* **74**, 6045.

NAME INDEX